# Towards Sustainability in the Wine Industry by Valorization of Waste Products

This volume in our Sustainability: Contributions through Science and Technology series reviews the use of alternative green technologies (pressurized liquid and supercritical fluid extractions) for grape biomass valorization. Environmental sustainability and circular economy are discussed in relation to agro-industrial waste in the winemaking industry. The waste contaminates water and soil and, in large quantities, it has been related to bad odors, a high content of organic matter in water, and greenhouse gas emissions over the entire wine-making industry. Here, the authors illustrate how green extraction of commercially valuable substances can be scaled up at an industrial level.

Features:

- Reports on waste valorization in the winemaking industry and converting the waste into more useful products including oils, antioxidants, and other valuable materials
- Explores research which contributes to environmental sustainability and circular economy in the winemaking industry
- Describes other ways to reduce the ecological footprint of the wine industry such as using less fertilizer, more benign pesticides, and reduction of water footprint
- Proposes options for a potential wine waste biorefining
- Reviews alternative uses of agro-industrial wine wastes as sources of additives for the food, cosmetic, and pharmaceutical industries

# Sustainability: Contributions through Science and Technology

Series Editor: Michael C. Cann, Ph.D.
*Professor of Chemistry and Co-Director of Environmental Science*
*University of Scranton, Pennsylvania*

## Preface to the Series

Sustainability is rapidly moving from the wings to center stage. Overconsumption of non-renewable and renewable resources, as well as the concomitant production of waste has brought the world to a crossroads. Green chemistry, along with other green sciences technologies, must play a leading role in bringing about a sustainable society. The Sustainability: Contributions through Science and Technology series focuses on the role science can play in developing technologies that lessen our environmental impact. This highly interdisciplinary series discusses significant and timely topics ranging from energy research to the implementation of sustainable technologies. Our intention is for scientists from a variety of disciplines to provide contributions that recognize how the development of green technologies affects the triple bottom line (society, economic, and environment). The series will be of interest to academics, researchers, professionals, business leaders, policy makers, and students, as well as individuals who want to know the basics of the science and technology of sustainability.

Michael C. Cann

**Green Organic Chemistry in Lecture and Laboratory** *Edited by Andrew P. Dicks, 2011*

**A Novel Green Treatment for Textiles: Plasma Treatment as a Sustainable Technology** *C. W. Kan, 2014*

**Environmentally Friendly Syntheses Using Ionic Liquids** *Edited by Jairton Dupont, Toshiyuki Itoh, Pedro Lozano, Sanjay V. Malhotra, 2015*

**Catalysis for Sustainability: Goals, Challenges, and Impacts** *Edited by Thomas P. Umile, 2015*

**Nanocellulose and Sustainability: Production, Properties, Applications, and Case Studies** *Edited by Koon-Yang Lee, 2017*

**Sustainability of Biomass through Bio-based Chemistry** *Edited by Valentin Popa, 2021*

*Nanotechnologies in Green Chemistry and Environmental Sustainability, 2022*

**Towards Sustainability in the Wine Industry by Valorization of Waste Products: Bioactive Extracts** *Edited by Patricia Joyce Pamela Zorro Mateus and Siby I. Garcés Polo, 2023*

# Towards Sustainability in the Wine Industry by Valorization of Waste Products

## Bioactive Extracts

Edited by
Patricia Joyce Pamela Zorro Mateus
and Siby I. Garcés Polo

CRC Press
Taylor & Francis Group
Boca Raton London New York

CRC Press is an imprint of the
Taylor & Francis Group, an **informa** business

First edition published 2023
by CRC Press
6000 Broken Sound Parkway NW, Suite 300, Boca Raton, FL 33487-2742

and by CRC Press
4 Park Square, Milton Park, Abingdon, Oxon, OX14 4RN

*CRC Press is an imprint of Taylor & Francis Group, LLC*

ISBN: 9781032465852 (hbk)
ISBN: 9781032489490 (pbk)
ISBN: 9781003391593 (ebk)

DOI: 10.1201/9781003391593

Typeset in Times
by Newgen Publishing UK

# Contents

# Preface

This volume in our Sustainability: Contributions through Science and Technology series reviews the use of alternative green technologies (pressurized liquid and supercritical fluid extractions) for grape biomass valorization. Environmental sustainability and circular economy are discussed in relation to agro-industrial waste in the wine-making industry. The waste contaminates water and soil and, in large quantities, it has been related to bad odors, a high content of organic matter in water, and greenhouse gas emissions over the entire wine-making industry. Here, the authors illustrate how green extraction of commercially valuable substances can be scaled up at an industrial level.

- Reports on waste valorization in the wine-making industry and converting the waste into more useful products including oils, antioxidants, and other valuable materials
- Explores research which contributes to environmental sustainability and circular economy in the wine-making industry
- Describes other ways to reduce the ecological footprint of the wine industry such as using less fertilizer, more benign pesticides, and reduction of water footprint
- Proposes options for a potential wine waste biorefining
- Reviews alternative uses of agro-industrial wine wastes as sources of additives for the food, cosmetic, and pharmaceutical industries

# Acknowledgments

The authors thank the Universidad Libre for financial support through the research project "Use and recovery of waste generated by the Colombian wine agro-industry by obtaining additives with potential application in the food, cosmetic and pharmaceutical industries using green technologies." We also thank the Universdiad Nacional de Colombia who provided support in the experimental development of the project.

# Editors

**Patricia Joyce Pamela Zorro Mateus** – Chemist and MSc. in Biochemistry from the National University of Colombia. I have academic experience in topics such as characterization of bacterial communities, population control by inhibition of quorum sensing between bacteria and, currently, in valorization of wastes. In this field I have been working in biodegradation of xenobiotics from fungi and bacteria to find degradation products that could be reused, and I also have been working with grape agro-industrial wastes to find natural oils, antioxidants, and biologically functional compounds. I have nine years of teaching experience; I currently work as a full-time chemistry professor at the Universidad Libre – Bogotá. In the past, I have also taught at the Jorge Tadeo Lozano University, Manuela Beltrán University, and at the Open and Distance University (UNAD), in which I designed the Natural Chemical Reactions course (module, laboratory guide and virtual course) for the Chemistry career.

**Siby I. Garcés Polo** – Chemical Engineering, B Sc., Msc and PhD from Universidad Pública de Navarra (Spain). Researcher and Research Chief of Faculty of Engineering in Universidad Libre Barranquilla. My experience is focused on gas adsorption processes for environment applications. In my last study, porous materials for $CO_2$ capture and catalysts for dry reforming reaction were developed and studied, respectively. In academic terms, I currently teach Environmental Management and Chemistry. I am also interested in studying wastewater treatment, valorization of biomass and pyrolysis.

# Contributors

**Angie Paola Toro Cardona**
Universidad Libre
Bogotá D.C., Colombia

**Alba Sofía Parra Carvajal**
Universidad Libre
Bogotá D.C., Colombia

**Jessica Tatiana Mancera Cifuentes**
Universidad Libre
Bogotá D.C., Colombia

**Liced Alejandra Basto Gómez**
Universidad Libre
Bogotá D.C., Colombia

**Karen Julieth Arce Jiménez**
Universidad Libre
Bogotá D.C., Colombia

**Ana María Cuta Martínez**
Universidad Libre
Bogotá D.C., Colombia

**Patricia Joyce Pamela Zorro Mateus**
Universidad Libre
Bogotá D.C., Colombia

**Oscar L. Ortiz Medina**
Universidad ECCI
Bogotá, Colombia

**María Alejandra Castañeda Muñoz**
Universidad Libre
Bogotá D.C., Colombia

**Jenny Viviana Bejarano Pérez**
Universidad Libre
Bogotá D.C., Colombia

**Siby I. Garcés Polo**
Universidad Libre
Barranquilla, Colombia

**Faride Geraldine Jiménez Rodríguez**
Universidad Libre
Bogotá D.C., Colombia

**Henry Isaac Castro Vargas**
Universidad Nacional de Colombia
Bogotá D.C., Colombia.
Thar Process Inc.
Pittsburgh PA, USA

**Daniela Méndez Velásquez**
Universidad Libre
Bogotá D.C., Colombia

# 1  The Environmental Issue of the Wine Industry Waste and Its Recovery
## An Overview

*Siby I. Garcés Polo, Oscar L. Ortiz Medina, and Patricia Joyce Pamela Zorro Mateus*

## ABOUT THE WASTE

Waste is any material in a solid, liquid, or gas state resulting from the process of extraction, transformation, manufacturing, or consumption where abandonment of the residue or waste by the owner is optional or mandatory. In relation to the wine-making industry, an estimated annual generation of between 2 and 3 million tons of waste or by-products is produced annually, mainly during the period of grape harvesting due to its stationary nature (Ventosa, 2011).

The wine production industry specifically generates solid waste that, taken together with the lignocellulosic waste, draws attention from researchers worldwide not only because of the possibility of reusing this type of waste in agriculture, but also because this solid waste represents a bio-source for the synthesis of new products once they are fractionated according to their chemical composition and properties (Pujol, 2013). The chemical composition of grape residues depends on various factors including, but not limited to, the geographic source, climate, cultivation time, and the variety of the grape (Pujol, 2013). Grape residues have also been investigated as a source of cellulose and hemicellulose (Spigno, 2008), natural antioxidants by the extraction of phenolic compounds (Garcia-Perez, 2010), fermentable sugars through enzymatic treatment for production of biofuels (Mazzaferro, 2011), fabrication of activated carbon (Deiana, 2009), and production of polyhydroxyalkanoates (Adriana Kovalcik, 2020), among others. Likewise, grape seed cannot only be used for propagating the crops, however, the seed has also been investigated for the extraction of bioactive substances (Yan Chena, 2020). Once some of the potential applications for grape residues used for wine production are known, ascertainment of the residues generated from the process is paramount.

One of the key processes in the making of wine is grape pressing to obtain its juice or fluid. The residues of this process include the stems, skins, and seeds, which are collectively known by the name pomace (Jelley, 2022) and these correspond to 62%

DOI: 10.1201/9781003391593-1

**TABLE 1.1**
**Wastes generated based on the production process**

| Process | Description of waste |
| --- | --- |
| Washing, cleaning, and mechanical reduction of raw matter | Grape pomace |
| Chemical treatment | Fermentation lees |
| | Bentonite; active carbon |
| Cleaning of process equipment and facilities | Sludges from on-site treatment of effluents |
| Storage and packaging | Paper and cardboard containers |
| | Glass containers |
| Addition of sodium hydroxide | Caustic soda containers |
| Fermentation and pressing | Agrochemical waste containing hazardous substances |
| Equipment maintenance | Synthetic hydraulic oils |

of the total residues from wine-making activities (Ruggieri et al., 2009). Likewise, there are other residues such as the lees (precipitates formed during wine making) that obtained during the wine fermentation clarification process which account for 14% of total residues; the stem (12%) made up of vine branches and leaves, and wastewater treatment sludge (12% of total waste). Although these are the most representative wastes in the wine sector in terms of processing grapes to obtain wine, each winery has other wastes associated with office, bottling, packaging, transportation, and facility maintenance activities. The main wastes generated and the activities or processes from which they are derived are detailed in Table 1.1.

The wine industry generates waste with a high content of organic matter, dyes, and water pH-reducing agents. Consequently, every effort must be directed to reduce, eliminate, or mitigate the contamination that may occur where final disposal of this type of waste is inappropriate. Among the main wastes or by-products from wine production, organic waste such as the pomace, seeds, stems, and leaves, corresponds to a weight of approximately 20% of processed grape total mass and is characterized by having high demand of chemical and biochemical oxygen (COD and BOD, respectively), whereby the residual organic matter is subject to the action of degrading microorganisms, thus generating greenhouse gases and harmful odors throughout the decomposition processes, therefore contaminating the receiving environment.

Based on the above, it must be understood that the generator of this type of waste material is required to hold liability for its transportation and final disposal, which also impacts the industry's finances. However, not everything is so negative, since grape pomace is a good source of phenolic compounds, a diverse family of secondary metabolites with unique chemical and biological properties whose main characteristic is their potential role as bioactive antioxidants. Phenolic compounds cannot be synthesized by any animal species. Consequently, they must be consumed naturally through the intake of certain foods or through nutritional supplements. The basic chemical structure of phenolic compounds in pomace comprises a hydroxyl group attached to an aromatic ring, thus allowing these compounds to interact with other chemical species through hydrogen bonds, esterification reactions, glycosylation, and hydrophobic interactions. These reactions are frequently associated with biological activities in the human body. In addition, these phenolic compounds are part of the substances that provide the organoleptic properties to food (Redagrícola, 2019).

## ENVIRONMENTAL ISSUES

The agriculture and food industries are among the most important in the industrial sector, not only in terms of economy but also in terms of environmental sustainability because of the high demand for resources. In this sense, the contribution of this sector to global warming ranges from 22% to 34% (Notarnicola, 2010); likewise, consumption of potable water by the agriculture and food industry is 70% (Bonamente, 2016).

Specifically, within the sector of agricultural and food products, the grape is one of the fruits most commonly cultivated worldwide, with a production of approximately 60 million tons per year, with about 80% of grape crops being used for wine production, and grape pomace representing about 20% of the mass of processed grape (Chand, 2009). Grape cultivation for wine-making processes requires some steps posing negative effects on the soil, water, and air. Therefore, such processes, as well as the residues from these processes, require special attention. In addition, if grape pomace is considered an abundant source of organic matter and a wide variety of valuable compounds, and with this the productivity of the agriculture and food sector in terms of economy and environmental impacts is open to improvement, there is thus a need to learn about the different options for their recovery and the economic, environmental, and social benefits to be highlighted.

Wine production has its beginning when the juice of these fruits is extracted. A sugar-rich liquor containing a high amount of other organic substances is collected, which yields the grape juice with its characteristic color, smell, and flavor. Grape pomace is the solid material left as residue after juice extraction. As previously stated, this material has a high content of organic matter and tannins, among other compounds, and most of these can be extracted and exploited for other processes of wine production or other industries. The management of these residues must be carried out properly to avoid contamination of the soil, water, and air. For soil and water, the organic matter and tannins may increase the demand for oxygen in these resources and therefore their ecological quality. On the other hand, decomposition of this organic matter and tannins by the action of microorganisms either under aerobic or anaerobic conditions leads to the transformation of solid material into gaseous compounds that potentiate the greenhouse effect, a causing agent of climate change, and harmful odor substances that disturb the environment and affect the health and well-being of people.

Pomace is an abundant residue in wine production and its management is potentially problematic since it has no obvious use. If pomace is not disposed of in sanitary landfills, the two most common methods for its beneficial use are as feed for animals and as soil fertilizer, but only a small part of these materials is used in these activities (Jelley, 2022). The nutritional value of grape pomace is relatively low, resulting in the use of only 3% for production of animal feed (Brenes, 2016). The foregoing implies that most pomace is disposed of in sanitary landfills or incinerated, hence pomace exploitation does not take place.

There are many activities of the agriculture industry that generate solid, liquid, and gas residues that very often become residues on which no adequate treatment is performed and that generate an environmental effect on different natural resources. An example of such environmental impact is air pollution caused by the emission of polluting gases or gases that cause the greenhouse effect. This type of pollution is currently analyzed in terms of the carbon footprint, which is the sum of greenhouse gas (GHG) emissions across several sources produced by a given human activity and conventionally expressed as kilograms of carbon dioxide equivalent ($CO_2$-eq). For wine production resulting from total or partial fermentation of fresh grapes or their must, complex biological and biochemical interactions between grapes and different microorganisms take place. Mainly, yeasts and lactic acid bacteria that generate $CO_2$ emissions into the atmosphere (Fleet, 2003) participate in grape fermentation. The carbon footprint allows the collection of substantial data from wine production on the environmental effect and may give clues about the complexity of the issue and its interrelationships on a global scale. An estimate based on 29 studies indicates that the global average carbon footprint generated by the production of 1 generic bottle of wine (0.75–1 L) over its entire life cycle (from the cradle to the grave) is 2.2 ± 1.3 kg of $CO_2$-eq (Rugani, 2013), which is comparable to driving a small gasoline car for about 20–30 kilometers. The involved processes considered for estimation of this carbon footprint included: vineyard planting, viticulture and grape growing, wine-making activities, packaging processes, transportation and distribution, storage and consumption, and end-of-life processes, with the viticulture activities (17%) and packaging processes (22%) being the most significant contributors to the carbon footprint. According to these data, the contribution to the human activity annual global carbon footprint around the world from the wine production sector can be estimated at approximately 0.3%, a value that should definitely not be overlooked (Rugani, 2013).

In summary, regarding the industrial wine-making sector, the environmental aspects are related to: (i) the use of water at cultivation stages and the winery processes, of which 70% becomes wastewater with acidic pH, high concentration of sulfides, sodium, and organic matter; (ii) solid, organic residues as a result of pruning activities at vineyards and those derived from the winery process such as pomace, seeds, lees, and dehydrated sludge, all these residues are managed through incineration processes, disposed of in landfills, or recovered. The inorganic residues are generated in winery processes and encompass packaging and cleaning processes and as an alternative to waste management,. recycling is considered (iii) The use of energy and emission of greenhouse gases involved in the entire process of production and post-production which through mechanisms of efficient use of energy and accountability for greenhouse gas emissions can be reduced and; (iv) use of chemical agents, especially the use of fertilizers, pesticides, and herbicides at the vineyards and in the winery processes for cleaning and disinfection procedures, washing of bottles that affect water resources through changes in the quality of wastewater, and a loss of soil fertility. Other related issues have to do with the use of land for grape crops and its effect on the ecosystems. Overall, the main environmental effects associated with vineyards and the winery processes are the eutrophication of fresh and sea water, water consumption, water and soil ecotoxicity, human toxicity, increased concentration of greenhouse gases, use of land, climate change, and the increasing impairment of the ozone layer (Gancedo, 2018).

## RECOVERY OF WASTE FROM THE WINE INDUSTRY AND ECONOMIC IMPACT

One of the commitments by the industry for contributing to the circular economy and sustainable production is giving added value to waste in order to increase the competitiveness of organizations, and also to reduce the environmental effects. The agriculture/food industry is one of the largest emitters of non-recovered waste, having an uncontrolled effect on the environment associated with a high demand for this type of products.

In particular, the wine industry is one of the most productive sectors worldwide (Devesa-Rey, 2011). According to estimates by the International Organization of Vine and Wine, wine production worldwide, excluding juices and musts, was 260 millions hectoliters (hL) in 2020, where production in European Union countries accounted for 165 million hL. For South America, the countries producing higher amounts are Argentina, Chile, and Brazil, which are also where greater consumption takes place (OIV, 2020). According to the figures above, the wine industry impact on the global economy yielding a revenue of USD 306.2 billion in 2021 (Statista, 2022).

Despite this worldwide production and revenue figures, the main concern is that the wine-making process generates a wide variety of solid and liquid by-products throughout the life cycle of the product, including vine shoots, grape pomace, wine lees obtained from decantation processes, filter residues, cakes, vinasses, and residual water generated by the wine-making lees and containing grape pulp, skins, seeds, and dead yeasts used for alcoholic fermentation (Devesa-Rey, 2011). Treatment, disposition, and adequate re-utilization of these waters is required to avoid negative environmental effects such as phytotoxic effects on crops (Moldes, 2008; Gómez-Brandón, 2019). It is estimated that for every 10 liters of wine produced, 3 kg of grape pomace, one of the main by-products of wine production, are generated (FEDNA, 2022). In this sense, recycling, recovery, and reduction of waste disposal have gained growing interest in the wine industry and scientific community.

Specifically, the recovery processes have covered several fronts, thus highlighting these residues as a path that may contribute to economic growth, especially since they are considered a source of natural antioxidants that are regarded as harmless when compared to the antioxidants synthetically collected and commonly used in economic terms.

Some of the recovery alternatives proposed for these residues include the following:

- Extraction of chemical and bioactive compounds: from grape pomace, which is mainly composed of grape skin and seeds, dietary fibers (soluble and insoluble) and phenolic compounds can be obtained from the grape skin which can be extracted and used as functional compounds for the food, pharmaceutical, and cosmetic industries due to their antioxidant and nutritional activity (Fontana, 2017; Kalli, Lappa, Bouchagier, et al., 2018; Lafka et al., 2007).

Other derivatives are pigments, tartaric acid, and lignocellulosic material in the form of polysaccharides (cellulose, pectin) and pomace and stem lignin. Different techniques are applied such as solvent extraction (methanol, acetone), microwave-assisted extraction, or green technologies and scalable technologies such as pressurized liquid extraction and supercritical $CO_2$ fluid extraction (Muhlack, 2018).

- Thermo-chemical conversion: grape pomace is a potential raw material to be used in biorefineries; however, the technology to be implemented for the best thermo-chemical conversion of this biomass should be tested in terms of economy, logistics for storage, and conversion efficacy. The technologies available for thermal conversion of dry biomass include: combustion, where energy created at the biomass oxidation process is usable for other processes or transformed into electricity; gasification to obtain syngas ($CO/H_2$) which can generate heat and electricity following combustion. In addition, liquid fuels can also be obtained through the Fischer-Tropsch reaction; and pyrolysis that produces a mixture of gas, liquid (bio-oil), and a solid (biochar) for energy or environmental applications (Muhlack, 2018).
- Composting and vermicomposting: a highly applied recovery pipeline is the grape pomace biodegradation process through composting under aerobic conditions for applications as a fertilizer; however, joint composting processes using organic waste from cities are recommended for neutralizing the pH drop due to the effect of the pomace physical-chemical characteristics. Likewise, vermicomposting provides a means to overcome this limitation (Gómez-Brandón, 2019).
- Feed for animals: in this case, grape pomace can be incorporated into the diet of ruminants to reduce digestibility. Also, the additional potential of recovery is in terms of mitigating the greenhouse effect since the content of tannins in grape pomace reduces the emissions of enteric methane (Jayanegara, 2012). Grape pomace supplementation in the diet for laying hens represents another application as it has been shown to improve the egg quality (Kara, 2015). For fattening pigs, inclusion of grape residues improved the growth performance of these animals and nutrient digestibility.
- Ethanol production: amongst the residues from wine production, grape pomace features a high content of soluble sugars that upon fermentation and in beverages such as schnapps (aguardiente) and residual sugar, provide industrial alcohol for cosmetic use. In this regard, the economic value of grape pomace is increased (Dávila 2019).

While the alternatives above have been extensively studied and tested, the greatest challenge is to harness the capacity to integrate the recovery pipelines and their technologies across the spectrum through larger frameworks such as biorefineries (Paini, 2022). However, for selecting the best recovery alternative with economic benefits, analysis of the carbon and water footprint also must be taken into account. In addition, the former considerations should not be overlooked for waste management processes.

A comparative evaluation of the recovery alternatives for grape pomace and biowaste treatment practices such as landfilling, anaerobic digestion, incineration, and composting has been carried out in terms of economy (operational costs) and carbon footprint ($CO_2$-eq). It was found that, for the carbon footprint, anaerobic digestion poses a highly unfavorable environmental performance due to direct emissions of GHGs such as methane. Nonetheless, the operative cost of around €12 per ton of waste is low. On the contrary, landfilling presents the lowest environmental loads from every alternative studied due to low GHG emissions, although the operating costs derived from landfill activities are the highest across the scenarios as obtaining an income from selling a product with a market value that allows reduction of operating costs is not possible. Second to landfilling, composting also exhibits poor economic results; however, the carbon footprint is comparable with the scenarios of landfilling and incineration, although values are lower as compared to the alternatives of anaerobic digestion and vermicomposting, whose poor unfavorable environmental results are due to the emissions of greenhouse gases. It is noteworthy that products with an added value (biogas, compost, vermicompost) are derived from these two processes leading to a subsequent improvement of the environmental profile. On the other hand, other methods such as incineration and vermicomposting are low-cost alternatives from an operative standpoint (Cortés, 2020).

This book is about the characterization of the waste from the agriculture-industrial processes of Isabella grape. In our studies, we investigate substances that could be functional for food and cosmetic industries. Supercritical fluids were used to extract these substances, mainly non-polluting materials such as water and $CO_2$. Our work contributes to the environmental sustainability and circular economy because it is aimed at adding value to the agriculture-industrial waste, which is usually discarded and ends up contaminating water and soil. In this work, Isabella grape is used by wine industries where grape pomace is discarded and is related to large disposed quantities of this material with bad odors, high contents of organic matter in water bodies and soil, and also linking it to greenhouse gas emissions over the entire chain of supply and production in the wine-making industry. The Isabella grape is a variety of grape widely used in Latin America to make wine; however, research work covering the potential use of Isabella grape waste is scarce.

The emerging extraction technologies used are recognized as environmentally friendly as they use GRAS solvents whose scaling potential is feasible and simple. Carbon dioxide ($CO_2$)-based supercritical fluid extraction (SFE) is recognized for its broad advantages and for having highly efficient industrial applications. This technology can provide high-quality, safe, and viable products for human or animal consumption. Currently, SFE has applications in the food, cosmetic, and pharmaceutical industries.

Finally, this book also offers a baseline study of an agriculture-industrial waste treated using fundamental techniques for the green extraction of commercially valuable substances that can be scaled up to the industrial level.

## REFERENCES

Adriana Kovalcik, I. P. (2020). Grape winery waste as a promising feedstock for the production of polyhydroxyalkanoates and other value-added products. Food and Bioproducts Processing, 1–10.

Bonamente, E. S. (2016). Environmental impact of an Italian wine bottle: Carbon and water footprint assessment. Science of the Total Environment, 560–561.

Brenes, A. V. (2016). Use of polyphenol-rich grape by products in monogastric nutrition. A review. Animal Feed Science and Technology, 1–17. doi:https://doi.org/10.1016/j.ani feedsci.2015.09.016

Chand, R. N. (2009). Grape waste as a biosorbent for removing Cr(VI) from aqueous solution. Journal of Hazardous Materials, 245–250. doi:https://doi.org/10.1016/j.jhaz mat.2008.06.084.

Cortés, A. M. (2020). Unraveling the environmental impacts of bioactive compounds and organic amendment from grape marc. Journal of Environmental Management, 111066. doi:https://doi.org/10.1016/j.jenvman.2020.111066

David Pujol, C. L. (2013). Chemical characterization of different granulometric fractions. Industrial Crops and Products, 494–500.

Dávila I., R. E. (2019). The Biorefinery Concept for the Industrial Valorization of Grape Processing By-Products. En C. M. Galanakis, Handbook of Grape Processing By-Products: Sustainable Solutions (págs. 29–53). London, UK: Academic Press, Elsevier. doi:https://doi.org/10.1016/B978-0-12-809870-7.00002-8

Deiana, A. S. (2009). Use of grape stalk, a waste of the viticulture industry, to obtain activated carbon. J. Hazard. Mater., 13–19.

Devesa-Rey, R. V.-A. (2011). Valorization of winery waste vs. the costs of not recycling. Waste Management, 31(11), 2327–2335. doi:https://doi.org/10.1016/j.wasman.2011.06.001

Díaz Sánchez, A. B. (2009). Reciclado del orujo de uva como medio sólido de fermentación para la producción de enzimas hidrolíticas de interés industrial. Universidad de Cádiz.

FEDNA. (22 de JUNIO de 2022). FEDNA. Obtenido de Fundación Española para el Desarrollo de la Nutrición Animal: http://fundacionfedna.org/ingredientes_para_pien sos/orujo-de-uva

Fleet, G. H. (2003). Yeast interactions and wine flavour. International Journal of Food Microbiology, 86(1–2), 11–22. https://doi.org/10.1016/S0168-1605(03)00245-9.

Fontana, A. R. (2017). Extraction, Characterization and Utilisation of Bioactive Compounds from Wine Industry Waste. En Q. Vuong (Ed.), Utilisation of Bioactive Compounds from Agricultural and Food Waste 1 (pág. 17). Boca Ratón: CRC Press. doi:https://doi.org/10.1201/9781315151540

Gancedo, A. S. (2018). Impactos ambientales derivados de la producción de vino de la D.O.P. Cangas. Trabajo fín de máster. Universidad de Oviedo.

Garcia-Perez, J. G.-A. (2010). Extraction kinetics modeling of antioxidants from grape stalk (Vitis vinifera var. Bobal): influence of drying conditions. J. Food Eng., 1–10.

Gómez-Brandón. M., H. I. (2019). Strategies for recycling and valorization of grape marc. Critical Reviews Biotechnology, 39(4), 437–450. doi:https://doi.org/10.1080/07388 551.2018.1555514

Jayanegara A., L. F. (2012). Meta-analysis of the relationship between dietary tannin level and methane formation in ruminants from in vivo and in vitro experiments. J. Anim. Nutr., 365–375.

Jelley, R. E.-B. (2022). First use of grape waste-derived building blocks to yield antimicrobial materials. Food Chemistry, 131025. doi:https://doi.org/10.1016/j.foodchem.2021. 131025

Kara, K. K. (2015). Effects of grape pomace supplementation to laying hen diet on performance, egg quality, egg lipid peroxidation and some biochemical parameters. Journal of Applied Animal Research, 303–310.

Kalli, E., Lappa, I., Bouchagier, P. et al. (2018). Novel application and industrial exploitation of winery by-products. Bioresour. Bioprocess. 5, 46. doi: https://doi.org/10.1186/s40 643-018-0232-6

Lafka, T. I., Sinanoglou, V., & Lazos, E. S. (2007). On the extraction and antioxidant activity of phenolic compounds from winery wastes. Food Chemistry, 104(3), 1206–1214. https://doi.org/10.1016/J.FOODCHEM.2007.01.068.

Mazzaferro, L. M. (2011). Production of xylooligosaccharides by chemo-enzymatic treatment of agricultural by-products. Bioresources, 5050–5061.

Moldes, A. V.-F. (2008). Negative effect of discharging vinification lees on soils. Bioresource Technology, 99(13), 5991–5996. doi:https://doi.org/10.1016/j.biortech.2007.10.004

Muhlack, R. A. (2018). Sustainable wineries through waste valorisation: A review of grape marc utilisation for value-added products. Waste Management, 72, 99–118.

Notarnicola, B. T. (2010). Including more technology in the production of a quality wine: the importance of functional. Proceedings of the 7th International Conference on LCA in the Agri-food Sector (pp. 235–240). BARI, Italy: Università degli Studi di Bari Settore Editoriale.

OIV. (2020). State of the world and vitivinicultural sector in 2020. International Organisation of Vine and Wine Intergovernmental Organisation. Obtenido de www.oiv.int/public/medias/7909/oiv-state-of-the-world-vitivinicultural-sector-in-2020.pdf

Paini, J. B. (2022). Valorization of Wastes from the Food Production Industry: A Review Towards an Integrated Agri-Food Processing Biorefinery. 13, 31–50. doi:https://doi.org/10.1007/s12649-021-01467-1

Redagrícola. (2019). Redagrícola. Obtenido de Redagrícola: www.redagricola.com/cl/los-residuos-vitivinicolas/

Rugani, B. V.-R. (2013). A comprehensive review of carbon footprint analysis as an extended environmental indicator in the wine sector. Journal of Cleaner Production, 54, 61–77. doi:https://doi.org/10.1016/j.jclepro.2013.04.036

Ruggieri, L., Cadena, E., Martínez-Blanco, J., Gasol, C. M., Rieradevall, J., Gabarrell, X., Gea, T., Sort, X., & Sánchez, A. (2009). Recovery of organic wastes in the Spanish wine industry. Technical, economic and environmental analyses of the composting process. Journal of Cleaner Production, 17(9), 830–838. https://doi.org/10.1016/J.JCLEPRO.2008.12.005.

Spigno, G. P. (2008). Cellulose and hemicelluloses recovery from grape stalks. Bioresource Technology, 4329–4337.

Statista. (2022). Statista. Obtenido de https://es.statista.com/estadisticas/1292549/ingresos-del-mercado-del-vino-en-el-mundo/

Ventosa, E.; Clemente, R.; Pereda, L. (2011). Gestión integral de residuos y análisis del ciclo de vida del sector vinícola: De residuos a productos de alto valor añadido. LifeHAproWINE.

Yan Chena, J. W. (2020). Effective utilization of food wastes: Bioactivity of grape seed extraction and its application in food industry. Journal of Functional Foods, 104–113.

# 2 Collection of Potentially Biologically Active Extracts from Isabella Grape Pomace Using Supercritical Carbon Dioxide

*Karen Julieth Arce Jiménez, Ana María Cuta Martínez, Patricia Joyce Pamela Zorro Mateus, and Siby I. Garcés Polo*

The wine industry in Colombia is positioned within the agro-industrial sector as one of the most important in terms of production due to the high consumption of supplies required for wine-making processes including large areas of soil occupied for cropping and harvesting grapes, the high consumption of water resources, and the use of agro-chemical products that contribute to the serial production of large quantities of wine. However, significant residues arise in these production processes which affect the quality of water, soil, and ecosystems. The production of wine implies both a high consumption of water resources as well as a significant production of wastewater containing grape residues, total suspended matter, and a high content of organic matter such as sugars, organic acids, and phenolic compounds, increasing the oxygen demand required for decomposition and degradation processes (Mossie et al., 2011; Gabzdylova et al., 2009).

Residues are mainly disposed of in surface water bodies or on soil, whereby through porosity the waste reaches groundwater bodies, thus altering the chemical composition of these elements and their stability as wastewater resulting from the wine-making industry is characterized by high salinity, high nitrate, sulfides, and organic matter content, as well as a low pH (Christ & Burritt, 2013).

The main by-products from grape milling processes are solid residues composed by brooms, grape pomace made out of peels, seeds, stems, and pulp (Cotacallapa Sucapuca et al., 2020). This waste builds up in large quantities, creating bad odors, a high content of organic matter in water bodies and soil, and is also linked to

DOI: 10.1201/9781003391593-2

greenhouse gas emissions over the entire wine-making industry supply and production chain (Christ & Burritt, 2013).

The grape residue, specifically the pomace, is also an important source of bioactive compounds as unfermented sugar, alcohol, and polyphenolic compounds such as flavonoids, anthocyanins, tannins, and hydroxycinnamic acids have been found in grape residue. The bioactive compounds above are known for their capacity as antioxidant, anti-inflammatory, anti-cancer agents, cell cycle regulators, and free radical scavengers (Mojica Gómez & Pérez Mora, 2019; Muhlack et al., 2017; Sirohi et al., 2020).

Currently, the management of solid waste in the wine agri-business is carried out through incineration or by organic matter composting practices (Mojica Gómez & Pérez Mora, 2019). However, the emerging residue exploitation techniques have proposed the collection of extracts containing bioactive compounds by means of extraction techniques including, but not limited to, supercritical fluid extraction, subcritical water extraction, ultrasound techniques, and high-pressure processes (Sirohi et al., 2020).

Different methodologies to evaluate the bioactive capacity of extracts collected from agro-industry residues are available. For example, antioxidant activity can be assessed from the application of synthetic free radicals. However, an emerging, increasingly implemented methodology utilizes yeast as a biological agent to analyze cell growth under conditions of induced oxidative stress. This eukaryotic organism also allows for the analysis of the main components in fruit residues resulting from industrial processes to promote cell growth and counteract cell oxidation (Moreno Gómez, 2014).

The objective of this study was to evaluate the capacity to exploit residues from the wine-making agri-business and the potential as antioxidant agents for protecting cell growth against oxidative stress. This study was aimed at analyzing the properties of the residues to be exploited for their qualities and essential components, in this case the *Vitis labrusca* L. grape residue which is the grape species mostly used in the Colombian wine-making industry using supercritical carbon dioxide ($CO_2$-SC) as extractant. This research allowed us to contribute to the body of knowledge on the behavior of this fruit, the contents of the grape, and the bioactive properties of the compounds in this residue, as well as its potential application as a nutritional supplement for properties such as cell protection and inhibition of oxidative stress.

## THEORETICAL FRAMEWORK

### FRUIT WASTE EXPLOITATION TECHNIQUES

The substantial wealth of effects mentioned above result from agro-industrial waste and such impact can be reduced by capitalizing on the use of residues, especially because of the content of bioactive compounds that can be widely used in different types of industries. The types of extraction facilitate obtaining these usable compounds and because of the different methods choosing the one that best suits according to the extraction matrix and the desired extract is possible.

In the study conducted by Benitez (2020) where the bioactive compounds in fruit residues such as mango leaves, olive leaves, grape pomace, sour orange peel, and sweet orange peel were analyzed through extraction techniques such as supercritical fluid extraction using supercritical $CO_2$, pressurized liquid extraction, and ultrasound-assisted extraction in which collected extracts were further tested for quality by analyzing total phenol using the Folin-Ciocaulteu test and spectrophotometry at 280 nm, and the antioxidant capacity by means of the DPPH test. This research resulted in ultrasound-assisted extraction having a higher performance; however, the other two types of extraction using high pressure and temperature are also functional in these cases (Beniítez Gil, 2020). On the other hand, the highest content of phenol was obtained from mango leaves followed by olive leaves, grape peels, and sour and sweet orange peels. Subsequently, by an antioxidant activity test, higher antioxidant activity was found in mango leaves. This level of antioxidant activity is followed, nonetheless, by that from grape peels, olive leaves, and sour and sweet orange peels. This result was very similar to that reflected in total phenol content expecting to have an antioxidant activity in the same order. However, grape peels have a higher potency as antioxidants compared to olive leaves considering that their phenol content is higher. A strong relationship between the total phenol content and antioxidant activity can be reported by this work as phenols are the potential bioactive compound responsible for the antioxidant activity (Beniítez Gil, 2020).

Grape residues from wine-making industries currently are the subject of research through the application of extraction techniques for by-products or value-added products such as polyphenols, a group of compounds with great capacity to maintain and improve brain function and enhance cognitive abilities. Fatty acids are also fruit compounds and have capabilities for cell protection and preventing hypertension and cardiovascular diseases (Aizpurua Olaizola et al., 2015). To evaluate the proportion of these compounds in grape residues from vineyards in Samaniego and Getaria, Spain, samples of red and white grape residues, peels, and seeds were prepared to obtain by application of supercritical fluids, polyphenolic compounds, and fatty acids. In this study, $CO_2$ in supercritical state was used and the variables pressure, temperature, and extraction time were optimized to obtain the grape non-polar fraction and fatty acids. Likewise, supercritical $CO_2$ was used with methanol as a co-solvent to obtain the grape polar compounds (polyphenols). Red grape was found to have a high content of fatty acids such as linoleic acid, with anti-inflammatory and anti-thrombotic potential, as well as oleic acid, a potential blood cholesterol regulator (Aizpurua Olaizola et al., 2015).

Agro-industrial waste has been studied from the implementation of extraction techniques using supercritical fluids since implementation of these procedures allows the collection of the low polar or non-polar extract or fraction with potential application in different industries. For example, passion fruit residue also has been studied by applying $CO_2$ as a solvent to elicit an oil with essential properties. The fatty acid composition of passion fruit oil is characterized by a high content of fatty acids such as linoleic, oleic, palmitic, stearic, linoleic, and palmitoleic acids, which are bioactive substances with potential applications in the cosmetic and pharmaceutical industries (Pantoja Chamorro et al., 2017).

For characterization of the potential properties of fruit residues other techniques such as classic extraction using co-solvents or ultrasound-assisted extraction have also been employed. This is the case of apple powder residues from the tea production industry in which analysis of phenolic compounds and fatty acids composition has also been reported by applying polar solvents such as water, methanol, and ethanol, whereby solubilization of phenolic compounds in the sample is possible. Likewise, the use of ultrasound extraction techniques was reported to collect fruit flavonoids that are mainly located in the cytoplasm and which can be diluted since structures and cell membranes can be broken down for compound solubilization by this technique. Agro-industrial residues from apple are also a source of flavonoids and phenolic compounds, hence they are featured as residues with high added value for potential use (Naffati et al., 2017).

## FRUIT RESIDUE ANTIOXIDANT ACTIVITY

Antioxidants are compounds capable of preventing cell oxidation in the body, which means that intracellular oxygen is decreased; this oxidation process is derived from the appearance of radicals and is linked to physiological aging and the onset of diseases (Vilaplana, 2007). Therefore, these compounds are important when it comes to prevent and treat pathologies and it is very important to widely include antioxidants in people's daily food ingest. The properties of antioxidants have generated a great deal of interest from consumers and the scientific community to further study where they are localized in different foods (Coronado et al., 2015).

Due to the above, research on the collection of antioxidants in fruit and vegetable residues (such as seeds, peels, branches, etc.) has been conducted since these are widely wasted and generate an environmental impact in their final disposition. For example, in a study of cantaloupe residues (seeds and peel) chemical and biochemical testing were carried out and important bioactive compounds were found. The compounds were tested on kidney, colon, and cervical cancer cells suggesting that with an antioxidant-rich diet the risk of contracting cancer may decrease. On the other hand, antioxidants can also be extracted from their source (fruits or vegetables) and used in the chemical, pharmaceutical, and food industries to benefit the products, thereby increasing their shelf life and maintaining quality (Rolim et al., 2018).

The physicochemical characterization and antioxidant activity of two types of fruits called *Physalis* were conversely analyzed as they have had a positive reception due to their flavor and potential medicinal use. The fruits were harvested and subsequently, when they were yellow, they were taken and preserved under three conditions: fresh, cold, and frozen. The total phenol content test used was according to the Follin-Ciocalteu method where the result was that a higher phenol content was obtained in the two samples at room temperature conditions, however, the phenol load decreases as the temperature drops meaning that refrigerated and frozen samples showed less phenol content, respectively. Antioxidant activity testing was by the DPPH method, obtaining a significant antioxidant activity from *Physalis pubescens* with respect to the other species. However, again the room temperature conditions were those displaying a representative antioxidant activity according to this method.

Finally, it was concluded that the conditions in which the fruits are stored are very important and although it is believed that freezing the fruits may prolong their life-time, the nutritional components may perhaps decrease after repeated freezing cycles, thus reducing quality (Grigolo et al., 2020).

## PHENOLIC COMPOUNDS' ANTIOXIDANT ACTIVITY

A large amount of waste having a considerable application potential is generated in agro-industries, such as the grape pisco industry, where like the wine industry seeds, peels, skins, among others, are disposed of as waste. In this case the result is a grape brandy from Peru and Chile. In a study conducted by Surco, Ayquipa, Quispe, García, and Valle (2020) the total amount of phenol and the antioxidant activity of eight types of grapes, *Vitis vinifera*, used for pisco production, were determined. The total phenol content was characterized with the Folin Ciocalteu test where values in mg of gallic acid/g seed are reported. Subsequently, the antioxidant activity was tested by a DPPH test consisting of the DPPH free radical uptake measured at 517 nm, and also with the FRAP test that is intended for testing the antioxidant compounds' redox power expressed in Trolox mM-equivalent (TEAC). As a result, it was found that the species with the highest TPC were Moscatel and Uvina, and those with the highest antioxi-dant activity were Uvina and Moscatel, which led the authors to conclude that the antioxidant activity is related to the total phenol content. However, the variety of phenols is wide and therefore differences in their values can be found, demonstrating that they are functional bioactive compounds (Surco Laos et al., 2020).

Recently, the extracts (bagasse and pulp) of four types of Amazonian berries (blue-berries, blackberry, willow, and aguaymant) were studied where antioxidant activity was tested by means of the DPPH assay discovering as a result that the antioxidant capacity in blueberry bagasse is the highest. Likewise, this fruit presented the highest content of phenols through the Follin-Ciocalteu test. This result shows that there may be a strong relationship of antioxidant activity with respect to the total phenol content as these contain a wide variety of compounds that exert this type of antioxidant cap-acity, among many other beneficial characteristics (Rojas Ocampo, 2020).

## YEAST ANTIOXIDANT ACTIVITY ASSAY

The antioxidant activity test in *Saccharomyces cerevisiae* cultures consists of indu-cing cell oxidation in this eukaryotic organism to analyze the antioxidant capacity of a matrix. The matrix used is usually from a natural source made out of fruit or plant extract, known for their high phenolic compounds content, which in turn is known for its cell protection properties against oxidative stress. In this assay the antioxi-dant activity of a matrix can be identified by means of the cell growth observed in yeast cultures once oxidative stress-induced conditions have been reached. The oxi-dizing agent used in these tests is typically $H_2O_2$, although studies using heavy metals as an oxidizing agent have been reported (Kerdsomboon et al., 2020). This assay uses a yeast culture as cells of this organism show similarities to human cells and their molecular behavior is very similar, and yeast is also a low-cost, easily cultivable organism (Moreno Gómcz, 2014).

Studies on species of the grape *Vitis vinífera* L., which is one of the grape types most widely used in the wine industry, have been reported where its ability to protect yeast cells from cell damage has been tested. It was found that grapes of the Merlot variety, specifically grape pomace extract, contain a high concentration of polyphenols, flavonols, and flavanols such as kaempferol, quercetin, and catechin, and that because of the composition it was one of the extracts displaying a greater capacity to promote yeast cell growth under conditions of $H_2O_2$-induced stress (Lingua et al., 2016).

## METHODOLOGY

### OBTAINING AND MAINTENANCE OF YEAST

Commercial, dry active bakery yeast (Levapan brand) was obtained and 1 g was added to 10 mL of sterile water. The solution was stirred and 100 μL were subsequently seeded in YPD agar medium. The solid-phase culture was incubated in a Petri dish at room temperature for 14 h and then yeast growth was confirmed.

### YEAST GROWTH CURVE

To report yeast growth, a yeast sample from the microorganism grown in YDP medium containing 1% yeast extract, 2% peptone, 2% glucose, 95% sterile water, and agar was taken using a circular loop. This medium was previously autoclave-sterilized for 15 min at 121°C and 103 kPa. After this, culture media were stored in a refrigerator at 4°C for 24 hours. Sterilization of the loop was carried out in dry heat using an alcohol burner and then the loop was placed in a Falcon tube with 8 mL of YDP. The loop-containing tube was allowed for growth overnight at 30°C and 150 rpm of agitation. After this time, a dilution in YDP broth was prepared whose 595 nm-absorbance reading yield was 0.100. For this measurement the Genesys 10S series UV-visible spectrophotometer was used. After harvesting a yeast culture of the expected optical density, 25 μL of this were added to 10 Falcon tubes containing 8 mL of YPD medium each. Falcon tubes were then shaken at 150 rpm and 30°C for 30 minutes. After this time the spectrophotometer was used to measure the absorbance for the 10 points sampled every 20 minutes, respectively.

### OBTAINING *VITIS LABRUSCA* L. GRAPE EXTRACTS AND TOTAL PHENOL CONTENT ANALYSIS

As mentioned in Chapter 3, grape residues (pomace) from the Casa Grajales wine company were used to obtain the grape extracts. The Isabella grape used as the matrix belongs to the species *Vitis labrusca* L. To prepare the sample, peels, stems, and seeds from the initial sample were selected and allowed to dry for 15 days in the absence of light. An extraction system was employed and supercritical carbon dioxide was used as solvent to obtain the polar substances from grapes. Extracts were obtained at different temperature and pressure conditions (20, 25, and 30 MPa, and 40, 50, and 60°C) and optimal conditions to obtain the extract with the highest content of functional compounds were calculated. All 12 extraction points were established and performed in triplicate. Quantification of the total phenol content was carried out by the Follin-Ciocalteu reagent test as described by Hosu et al. (2014). The extracts were measured on a spectrophotometer at a wavelength of 765 nm, the Follin reagent was used, and results were expressed as mg of gallic acid equivalent (mg GAE/g extract) (Hosu et al., 2014).

## OXIDATIVE STRESS ASSAY IN SACCHAROMYCES CEREVISIAE

The objective of this test is to evaluate the antioxidant capacity of grape extracts to promote growth of yeast cultures and compare the growth curve from the results of a positive control that contains ascorbic acid and hydrogen peroxide and a hydrogen peroxide negative control. The procedure was carried out based on the methodology described by Moreno (2014).

## ANTIOXIDANT ACTIVITY ASSAY CURVE

Plotting of the antioxidant activity resulted from preparation of a yeast dilution as mentioned above, allowing the diluted yeast to incubate in liquid medium overnight on a shaker at 150 rpm and 30°C. An optical density of 0.100 was achieved. From that yeast dilution, 25 µL were then added into 25 mL Falcon tubes. Each tube contained 8 mL of YDP medium. Positive controls were set up using ascorbic acid and hydrogen peroxide, a negative control using peroxide, a blank with yeast, and finally the grape extract sample at the quantities shown in Table 2.1. The testing extract item was that obtained under conditions of 32.1 MPa and 50°C with a total phenol content of 7.05 mg GAE/g extract representing the point where the highest total phenol content was collected. They were then incubated at 30°C and stirred at 150 rpm for 30 minutes. Measurements were taken every 30 min on a Genesys 10S series spectrophotometer at a 595 nm wavelength. Extract- and yeast-containing samples were analyzed in triplicate for each measurement. Measurements were performed over 180 minutes and each measurement was recorded every 30 minutes on the spectrophotometer. Likewise, manual stirring at each point was carried out before taking each

## TABLE 2.1
### Contents of tests to measure the antioxidant activity of extracts

|  | Blank | Positive control | Negative control | Sample |
|---|---|---|---|---|
| Culture medium (YDP) | 8.240 mL | 8.080 mL | 8 mL | 8 mL |
| Yeast | 25 µL | 25 µL | 25 µL | 25 µL |
| Ascorbic acid | – | 80 µL | – | – |
| Grape extract | – | – | – | 80 µL |
| $H_2O_2$ | – | 160 µL | 160 µL | 160 µL |
| Final volume | 8.265 mL | 8.265 mL | 8.265 mL | 8.265 mL |

*Note:* Taken from Nutritional analysis and study of the antioxidant activity of some tropical fruits grown in Colombia, by E. Moreno Gómez, 2014, final work of Master's Degree [Universidad Nacional de Colombia at Bogotá]. https://repositorio.unal.edu.co/handle/unal/53992.

measurement. The test was performed by analyzing each condition and all points were served under sterile, laminar-flow cabin conditions.

## RESULTS

### Yeast Growth Curve

The growth curve (Figure 2.1) was obtained to observe the metabolism and growth of bakery yeast and identify the curve fragments displaying a more accelerated growth. Absorbances at 60 minutes yielded values corresponding to a larger curve slope regarding other segments of the entire latency phase. At the measurement time of each point no signs of stationary phase or death phase were observed. However, a mild resemblance to exponential growth phase can be observed in points 4 through 6.

### Antioxidant Activity Assay Curve

Figure 2.2 shows the results for grape extract antioxidant activity. A constant growth of yeast with extract was observed throughout the data collection period. Growth was rapid within 40 minutes of measurement as in those fragments a larger slope curve is observed. The positive control had a steady growth during the first 2 hours of measurement, however, this growth was not observed to reach the cell concentration values obtained from the curve for yeast plus extract. Absorbance data showed that the grape extract promoted yeast cell growth through maintained, exponential growth as evidenced in the yeast growth curve. The negative control result showed that cell concentration within the first 40 minutes is close to zero; however, from the third

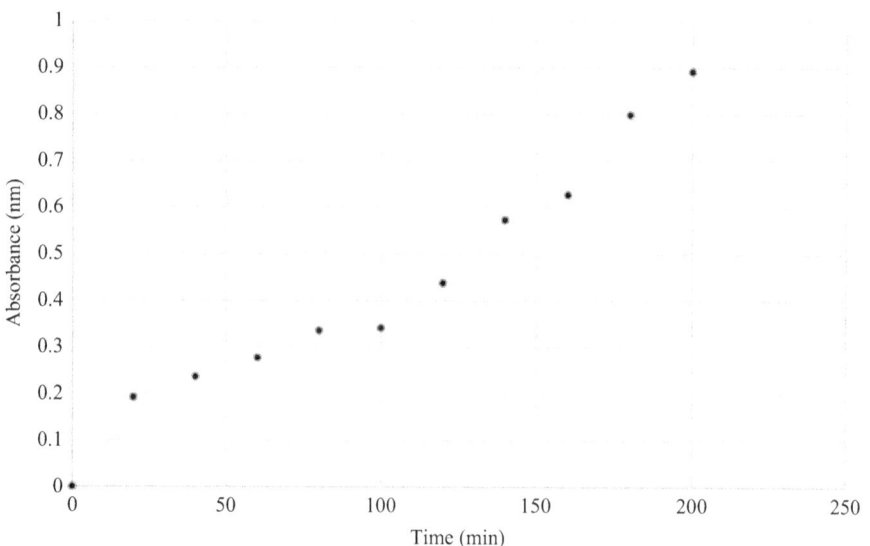

**FIGURE 2.1** *Saccharomyces cerevisiae* growth curve. Absorbances measured at 595 nm.

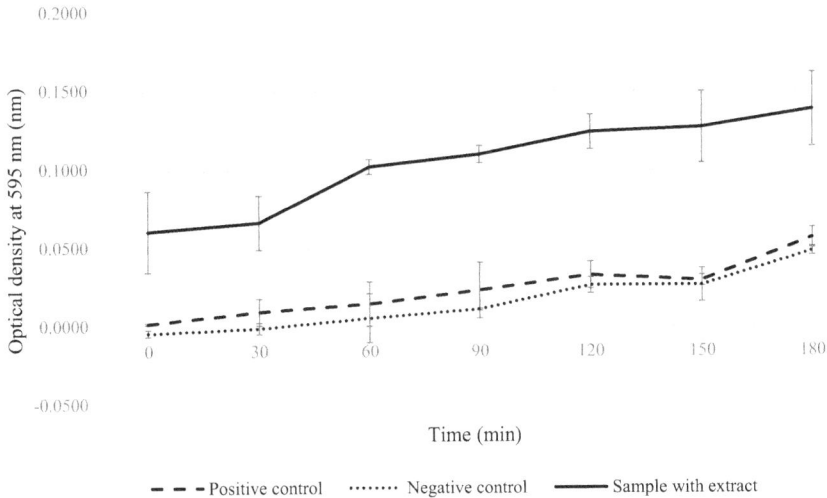

**FIGURE 2.2** Antioxidant activity of grape pomace extracts. The standard deviation is displayed as error bars.

measurement, absorbance values displaying discrete yeast growth in the presence of the oxidizing agent were obtained.

## DISCUSSION

The content of bioactive compounds in fruits constitutes a source of nutrients essential for cell protection, prevention of the appearance of cancer cells in the body, and the prevention of cardiovascular and neurodegenerative diseases (Creus, 2004). The listings with classifications and types of polyphenolic and phenolic compounds with broad bioactive properties are extensive. For example, the compounds with higher antioxidant capabilities among phenolic acids are p-coumaric acid, caffeic acid, ferulic acid, and ellagic acid (Pérez et al., 2015). Among the groups with higher antioxidant capacity are the flavonols due to their high antioxidant capacity and ability as donors of hydroxyl groups, which allows them to play a role in collagen stabilization of arterial walls, decreasing the inflammatory process in atheromatous plaque formations, and diminished oxidation of low-density lipoproteins (Creus, 2004). These compounds are concentrated in certain parts of the grape. Seeds, for example, in *Vitis labrusca* L. grapes contain the highest content of phenolic compounds (91.53–104.19 mg GAE/g sample on a dry basis), followed by peels and branches (Ruales Salcedo et al., 2017). In this study, grape pomace that contains seeds and peels was used due to the high content of antioxidant compounds and these were considered to be the parts of the residue having the greatest potential for use according to reported data.

In other studies, grapes of the species *Vitis labrusca* L. have been reported to possess a high content of polyphenols, phenols, and flavonols (mainly responsible for antioxidant activity) since grapes of this species have a total content of flavonols of 295.7 mg/100 g of grapes in dry weight, which represents in addition a two-fold total flavonoids content as compared to grape *Vitis vinifera* species such as Chardonnay, Sauvignon Blanc Vermentino, and Vioginer. The *Vitis labrusca* L. grape is also one of the species with the higher caffeic acid, P-coumaric acid, and catechin contents (Burin et al., 2014).

Furthermore, in the previous study, a value of 56.6 mg GAE/100 g of total phenol content was obtained which had an antioxidant capacity in the DPPH and ABTS assays of 104.3 and 208.6 µmol Trolox equivalents/100 g of sample that may be associated with antioxidant capacities of the sample used in this study as the optimal value estimated by the Statgraphic software was 53.47 mg GAE/100 g, which is very close to the values recorded in the literature. However, under the conditions closest to the optimum point a total phenol content of 39.83 mg GAE/100 g was obtained, representing the timepoint with the highest total phenol content. In this study, that antioxidant capacity of yeast was tested to determine the capacity of the extract to promote cell growth and increase cell density in yeast dilutions. However, comparison of results from this analysis with other studies is not yet possible as in yeast assay databases there is no record available for using *Vitis labrusca* L. grape extract as a natural antioxidant so far. It was possible however to determine that the antioxidant capacity of this grape is high due to the high content of phenolic compounds and testing conducted in other studies. Likewise, these qualities endow it with the property to become a value-added by-product in the industry which constitutes a natural source of bioactive compounds and whose exploitation is usable for replacing the current use of synthetic antioxidants such as butylhydroxyanisole (BHA), butylhydroxytoluene (BHT), and propyl gallate (GP), the additives most commonly used in industry that represents a high content of toxicity (Latorre Leal, 2016).

## CONCLUSIONS

This research found antioxidant substances in extract from Isabella grape pomace that were capable of counteracting the peroxide hydrogen effect, as an oxidant, in a test with *Saccharomyces cerevisiae*, a microorganism that was able to grow with a stable rate in the time period tested. These extracts were subjected to a total phenol content test detailed in Chapter 3, where evidence is reviewed as to one of the major characteristics of this antioxidant activity test, which is the concentration of extract phenols such as caffeic acid, P-coumaric acid, resveratrol, flavanols like catechin, and flavonols such as quercetin present in the *Vitis labrusca* L. grape (Burin et al., 2014; Pérez et al., 2015). Likewise, antioxidant activity may also be related to the presence of fatty acids in grape residues, such as oleic acid and linoleic acid (Kovlcik et al., 2020). Supercritical fluid extraction carried out on *Vitis labrusca* L. grape pomace allowed the collection of extracts containing potentially bioactive compounds that may have applications in the pharmaceutical, cosmetic, and food industries.

## GLOSSARY

**Absorbance:** Amount of light absorbed by a sample that depends on its concentration. Absorbance is defined by the negative logarithm of the intensity quotient after absorbing the light and the intensity of the light being sampled.

**Aging:** Due to the oxidizing action of free radicals. Therefore, antioxidants can be administered to decrease the effects of aging on the body. Approximately 2% of the oxygen used by cells is not converted into water but into reactive oxygen species. Most of these species originate in the mitochondria hence the importance of this organelle and, especially, of mitochondrial DNA for understanding aging.

**Agro-industry:** Economic sector defined as being made up of different types of industries linked to agricultural activities which include processing and marketing agricultural products. The food agro-industry that transforms food into different formats and properties thus becomes an industry. In addition, this economic sector usually is a consumer and supplier of used raw material.

**Antioxidant activity:** Ability of a substance to act against the oxidative stress of a cell as the antioxidant reacts with free radicals or reactive oxygen substances and protects the cell from aging or the formation of cancer cells. These antioxidants are naturally found in fruits, vegetables, and plants.

**Bakery yeast:** Yeast is a generic name that groups a variety of fungi including both plant and animal pathogenic species, as well as species that are not only harmless but very useful. Within the genus *Saccharomyces*, the species *cerevisiae* constitutes the yeast that is the most studied eukaryotic microorganism. This organism is also known as bakery yeast, since adding the yeast to the dough used to prepare bread causes it to rise as required. From a scientific point of view, studying yeasts as a biological model has greatly contributed to elucidating the basic processes of cell physiology.

**Bioactive compounds:** Components of fruits, vegetables, nuts, oils, beans, and natural plants that have the ability to interact in body tissues through reactions that improve human health, prevent diseases, and improve the physical and mental state of humans.

**Extract:** Substances obtained through the use of solvents, arranged in close contact with the main solid sample or substance to obtain or extract the active compounds. These substances with bioactive properties often have components that may be beneficial for investigation or to improve existing products.

**Free radicals:** Free radicals are those molecules that have an unpaired electron in their outermost orbital. This gives them a very high reaction capacity so they can act producing changes in the chemical composition or structure of cell elements that make them incompatible with life in biological systems.

**Growth curve:** A data modeling tool to analyze the growth statistics of a population, sample, study purpose which serves to identify the growth metabolism over time to be able to achieve the different growth phases or identify the stationary phase when growth is halted.

**Isabella grape:** A shrub whose scientific name is *Vitis labrusca* L., a species of grape sown mainly in the Valle del Cauca, a region of Colombia where the area of cultivation for this grape is greater. Isabella grape is classified as Vines in the upper classification of the species and belongs to the class Magnoliopsida's *Vitaceae* species, and is widely used in red wine-making.

**Oxidative stress:** Process where intracellular free radicals are increased and not eliminated by antioxidants, which leads to an increased intracellular oxidative activity causing tissue deterioration and favoring the onset of diseases such as cancer, cardiovascular diseases, skin premature aging, neurological disorders, among other diseases.

**Phenolic compounds:** Substances that have hydroxybenzene functions and are linked to aromatic or aliphatic substances. Phenolic compounds also encompass a group of micronutrients from the plant kingdom that have anti-inflammatory and anticancer properties, and physiological and metabolic effects on the human body.

**Spectrophotometer:** Laboratory apparatus where a sample or solution is exposed to a white light source and then the light reflected through some light intervals is calculated. The main objective of using this tool is learning the concentration of the solution.

## REFERENCES

Aizpurua Olaizola, O., Ormazabal, M., Vallejo, A., Olivares, M., Navarro, P., Etxebarria, N., & Usobiaga, A. (2015). Optimization of Supercritical Fluid Consecutive Extractions of Fatty Acids and Polyphenols from Vitis Vinifera Grape Wastes. Journal of Food Science, 80(1), E101–E107. https://doi.org/10.1111/1750-3841.12715

Beniítez Gil, P. (2020). Estudio del proceso de producción de sustancias bioactivas a partir de la fermentación en estado sólido de residuos agroalimentarios y su extracción a alta presión. Trabajo de grado. [Universidad de Cádiz, Cádiz]. http://hdl.handle.net/10498/23692

Burin, V. M., Ferreira Lima, N. E., Panceri, C. P., & Bordignon Luiz, M. T. (2014). Bioactive compounds and antioxidant activity of Vitis vinifera and *Vitis labrusca* grapes: Evaluation of different extraction methods. Microchemical Journal, 114, 155–163.https://doi.org/10.1016/j.microc.2013.12.014

Christ, K. L., & Burritt, R. L. (2013). Critical environmental concerns in wine production: An integrative review. Journal of Cleaner Production, 53, 232–242. https://doi.org/10.1016/j.jclepro.2013.04.007

Coronado H, Marta., Vega, Salvador., Gutiérrez T, León. Rey,. Vázquez F, Marcela & Radilla V, Claudia. (2015). Antioxidantes: perspectiva actual para la salud humana. Revista Chilena de Nutrición, 0717–7518. 42(2) dx.doi.org/10.4067/S0717-75182015000200014

Cotacallapa Sucapuca, M., Vilca Curo, R., & Coaguila, M. (2020). El orujo de Uva Italia como fuente de compuestos bioactivos y su aprovechamiento en la obtención de etanol y compost. FAVE – Ciencias Agrarias, 19(1), 17–32. https://doi.org/10.14409/fa.v19i1.9450

Creus, E. G. (2004). Compuestos fenólicos. Offarm, 6(10), 79–86. www.elsevier.es/es-revista-offarm-4-articulo-compuestos-fenolicos-un-analisis-sus-13063508

Gabzdylova, Barbora., Raffensperger, John. F., Castka, Pavel. B. (2009). Sustainability in the New Zealand wine industry: drivers, stakeholders and practives. Journal of Cleaner Production, 17(11), 992–998. doi:10.1016/j.jclepro.2009.02.015

Grigolo, C. R., Oliveira, M. D. C., Loss, E. S., Ropelato, J., Oldoni, T., & Batista Lafay, C. B. (2020). Caracterización fisicoquímica y contenido antioxidante de frutas de Physalis. Revista Mexicana de Ciencias Agrícolas, 11(3), 607–618. https://doi.org/10.29312/remexca.v11i3.2080

Hosu, A., Cristea, V. M., & Cimpoiu, C. (2014). Analysis of total phenolic, flavonoids, anthocyanins and tannins content in Romanian red wines: Prediction of antioxidant activities and classification of wines using artificial neural networks. Food Chemistry, 150, 113–118. https://doi.org/10.1016/j.foodchem.2013.10.153

Kerdsomboon, K., Chumsawat, W., & Auesukaree, C. (2020). Effects of Moringa oleifera leaf extracts and its bioactive compound gallic acid on reducing toxicities of heavy metals and metalloid in Saccharomyces cerevisiae. Chemosphere, 1–10. https://doi.org/10.1016/j.chemosphere.2020.128659

Kovalcik, A., Pernicova, I., Obruca, S., Szotkowski, M., Enev, V., Kalina, M., & Marova, I. (2020). Grape winery waste as a promising feedstock for the production of polyhydroxyalkanoates and other value-added products. Food and Bioproducts Processing. 124, 1–10. doi:10.1016/j.fbp.2020.08.003

Latorre Leal, M. (2016). Polifenoles de la uva. Trabajo fin de grado [Universidad Complutense, Madrid]. https://eprints.ucm.es/id/eprint/49802/

Lingua, M. S., Fabani, M. P., Wunderlin, D. A., & Baroni, M. V. (2016). In vivo antioxidant activity of grape, pomace and wine from three red varieties grown in Argentina: Its relationship to phenolic profile. Journal of Functional Foods, 20, 332–345. https://doi.org/10.1016/j.jff.2015.10.034

Mojica Gómez, J., & Pérez Mora, W. (2019). Aprovechamiento de Residuos Agroindustriales de la Industria Vinicola del Valle de Sáchica (J. Mojica Gómez & W. Pérez Mora (eds.); 1st ed.). Servicio Nacional de Aprendizaje SENA. https://hdl.handle.net/11404/5389

Moreno Gómez, E. (2014). Análisis nutricional y estudio de la actividad antioxidante de algunas frutas tropicales cultivadas en Colombia. Trabajo final de Maestría [Universidad Nacional de Colombia, Bogotá]. https://repositorio.unal.edu.co/handle/unal/53992

Mossie, K.P.M., Patti, A.F H., Christen, E. W., Cavagnaro, T. R. (2011). Review: Winery wastewater quality and treatment options in Australia. Australian Journal of Grape and Wine Research, 17(2), 111–122. doi:10.1111/j.1755-0238.2011.00132.x

Muhlack, R. A., Potumarthi, R., & Jeffery, D. W. (2017). Sustainable wineries through waste valorisation: A review of grape pomace utilisation for value-added products. Waste Management, 72, 99–118. https://doi.org/10.1016/j.wasman.2017.11.011

Naffati, A., Vladić, J., Pavlić, B., & Vidović, S. (2017). Biorefining of filter tea factory by-products: Classical and ultrasound-assisted extraction of bioactive compounds from wild apple fruit dust. Journal of Food Process Engineering, e12572, 1–11. https://doi.org/10.1111/jfpe.12572

Pantoja Chamorro, A. L., Hurtado Benavides, A. M., & Martinez Correa, H. A. (2017). Caracterización de aceite de semillas de maracuyá (*Passiflora edulis* Sims.) procedentes de residuos agroindustriales obtenido con CO2 supercrítico. Acta Agronomica, 66(2), 178–185. https://doi.org/10.15446/acag.v66n2.57786

Pérez, C., Ruiz Del Castillo, M. L., Gil, C., Blanch, G. P., & Flores, G. (2015). Supercritical fluid extraction of grape seeds: Extract chemical composition, antioxidant activity and inhibition of nitrite production in LPS-stimulated Raw 264.7 cells. Food and Function, 6(8), 2607–2613. https://doi.org/10.1039/c5fo00325c

Rojas Ocampo, E. (2020). Compuestos fenólicos, Actividad antioxidante y actividad antimicrobiana sobre levadura (*Saccharomyces cerevisiae*) del extracto de cuatro berries provenientes de la región de Amazonas. Tesis de grado. Universidad Nacional Toribio Rodríguez de Mendoza de Amazonas, Chachapoyas.

Rolim, P. M., Fidelis, G. P., Padilha, C. E. A., Santos, E. S., Rocha, H. A. O., & Macedo, G. R. (2018). Phenolic profile and antioxidant activity from peels and seeds of melon (*Cucumis melo* L. var. *reticulatus*) and their antiproliferative effect in cancer cells. Brazilian Journal of Medical and Biological Research, 52(e6969). https://doi.org/10.1590/1414-431X20176069

Ruales Salcedo, A. V., Rojas González, A. F., & Cardona Alzate, C. A. (2017). Obtención de compuestos fenólicos a partir de residuos de uva isabella (*Vitis labrusca*). Biotecnología

En El Sector Agropecuario y Agroindustrial, 15(Edición Especial 2), 72–79. https://doi.org/10.18684/bsaa(v15)ediciónespecial.580

Sirohi, R., Tarafdar, A., Singh, S., Negi, T., Gaur, V. K., Gnansounou, E., & Bharathiraja, B. (2020). Green processing and biotechnological potential of grape pomace: Current trends and opportunities for sustainable biorefinery. Bioresource Technology, 314(123771), 1–50. https://doi.org/10.1016/j.biortech.2020.123771

Surco Laos, F., Ayquipa Paucar, H., Quispe Gamboa, W., García Ceccarelli, J., & Valle Campo, M. (2020). Determinacion de polifenoles totales y actividad antioxidante de extracto de semillas de uvas de residuos de la producción de piscos. Revista de La Sociedad Química Del Perú, 80(2), 123–131. https://doi.org/10.37761/rsqp.v86i2.282

Vilaplana, Montse. (2007). Antioxidantes presentes en los alimentos. Vitaminas, minerales y suplementos. Offarm, 53, 232–242. www.elsevier.es/es-revista-offarm-4-articulo-antioxidantes-presentes-los-alimentos-vitaminas-13112893

# 3 Obtaining Potentially Functional Oils from Isabella Grape (*Vitis labrusca* L.) Pomace Using Supercritical Carbon Dioxide

*Alba Sofía Parra Carvajal, Angie Paola Toro Cardona, Patricia Joyce Pamela Zorro Mateus, Henry Isaac Castro Vargas, and Siby I. Garcés Polo*

The production of agro-industrial waste is currently a global problem. Therefore, it is important to implement waste management improvement strategies in the agri-food sector, such as recycling or a circular economy since, from an economic perspective, one of the advantages of reusing waste in this sector is the improvement of waste biomass for the generation of products that are sustainable and renewable for consumers thus contributing to protection of the environment (Campalani et al., 2020). For this reason, supercritical fluid extraction (SFE), considered an environmentally friendly extraction method, was implemented where the attraction and grouping characteristics of a molecule in a fluid above its critical point are studied. Supercritical fluid extraction is a very useful method for analyte separation from a certain sample (Hernández Romero, 2017).

Organic solvents such as $CO_2$, methanol, ethanol, water, methane, ethane, propane, ethylene, acetylene, among others are used in SFE (Rivera et al., 2016). This technique is based on using substances found in certain conditions such as gas or liquid according to pressure and temperature values above the critical point. In such a way, $CO_2$ is the solvent most used because its critical temperature is low with controlled critical pressure, and upon temperature variation $CO_2$ becomes a good solvent, thereby rendering this technique a cleaner and safer method compared to conventional extraction techniques (Hernández Romero, 2017).

Supercritical fluid extraction has been used in the production of various products from multiple plant materials. Among these are roasted coffee beans, grape seeds

DOI: 10.1201/9781003391593-3

such as pulp and peels, guava leaves, and others (Sovovà & Stateva, 2011). That is why this method is used in some applications, for example, in palm oil extraction, being a novel process to obtain residual oil from pressed palm fiber because it is one of the most abundant sources of carotene and vitamin A, as well as the extraction of different spices or fruits such as guava, papaya, guanabana, among others, extraction of which is performed using supercritical $CO_2$ that provides a high use profile in industries such as food, cosmetics, and pharmaceuticals due to multiple properties as antioxidant agents (Akanda et al., 2012). On the other hand, SFE has become an alternative to the generation and development of new technologies for extraction and purification of pollution in food and cosmetic industries, allowing to fundamentally modify many petroleum extractions and solid waste treatment at the technology level where SFE is a method seeking to have a more environment-friendly impact (Manjare & Dhingra, 2019).

As for the agri-food sector and particularly the wine-making sector, 20% of solid waste comes from grape processing where the disposed material includes peels, stems, and seeds from the cleaning, washing, and cutting steps. That is why the agri-food and wine-making sectors are mainly responsible for the environmental impact from human activities such as the use and quality of water. In the cropping step, large quantities of water are used and become wastewater because with no adequate treatment they cause eutrophication problems. Similarly, in regard to the use of energy and the emission of greenhouse gases by the wine sector, particularly at post-production phases such as bottling, packaging, and distribution, significant amounts of energy are consumed and therefore, a substantial amount of greenhouse gases is emitted (Gancedo Alonso, 2018). Thus, the importance of SFE in developments for treating grape residues lies in a reduction in the amount of residues generated with economic feasibility, because grape residues are an important source of commercial interest due to value-added properties (Vardanega et al., 2015).

Due to the aforementioned, this study sought to implement SFE using $CO_2$ as a solvent to obtain extracts from Isabella grape (*Vitis labrusca* L) residues since in Colombia production and demand for Isabella grapes has increased as the grape is a fruit with excellent sources of bioactive substances such as phenolic compounds, which have great commercial value (Ruales & Alzate, 2017).

To carry out this extraction, performing the optimization of factors such as temperature and pressure was decided to determine the extraction performance, total content of phenolic compounds (TPC), and the fatty acid profile.

## THEORETICAL FRAMEWORK

### SUPERCRITICAL FLUID (SCF)

An SCF is a liquid that is above its critical value at a certain temperature (Tc) and pressure (Pc) and has properties similar to gases and liquids. Among the physical-chemical characteristics of a SCF, high diffusivity and low viscosity can be found which provide the SCF solvent capabilities. Likewise, a SCF behaves better than a liquid due to the yield values being better because it is easily spreadable on solid material, and so the solubility of as SCF can be modified by increasing or decreasing the extraction pressure which increases the solvent capabilities (da Silva et al., 2016; Herrero et al., 2016).

## SUPERCRITICAL FLUID EXTRACTION (SFE)

Supercritical fluid extraction is a technique where analytes are separated from a solid sample or matrix by means of an extractive phase used as a solvent (Vargas Reyes, 2018). For this, the sample must be subjected to a treatment prior to the extraction operation.

The SFE process has several characteristics such as: high extraction efficiency, simple extraction technology, and no requirement for any other solvent recovery equipment. The importance of the SFE process lies in the numerous applications that have been established in industries such as the chemical industry, managing to yield large increases in added value to waste of plant origin through large-scale extractions for the agro-industry overall (Bhusnure et al., 2016).

There are several parameters that interfere with the extraction process: (i) Pressure influences the density since pressure determines the ability of a solvent to dilute the analyte at a constant temperature; (ii) temperature, which increases mass transfer by raising the analyte vapor pressure (in supercritical water at high temperatures near the critical point, increasing extraction performance and speed); (iii) the proportion of solute used, which influences the operation costs and also the amount of extract obtained; and (iv) pretreatment of the solid sample (i.e., the process of plant biomass drying, separation, and milling which may favor the extraction process) (Brunner, 2014). The SFE is developed in two important steps: (a) compounds found in the solid matrix are solubilized and separated by solvent carryover properties, a process that continues to be repeated as the solvent flows through the matrix and (b) the solvent and extract are separated due to temperature and pressure conditions since the solvent is released into the atmosphere (da Silva et al., 2016).

## SOLVENTS

In SFE, a green solvent technology is used as its potential is based on the density and varies by gradual increases either in pressure or temperature, which contribute to an increased extraction efficiency (Rivera Narváez, Carlos M.; Cardona Bermúdez, Liliana M.; Muñoz, Laura M.; Gómez, Dorely D.; Passaro Carvalho, Catarina; Quinceno Rico, 2016). Among the solvents used a great diversity can be found in Table 3.1.

Among the solvents shown in Table 3.1, ethanol is notable for being harmless and applied to obtain compounds for use in the food, cosmetic, and pharmaceutical industries (Rivera et al., 2016). There are two solvents with high potential, $CO_2$ and supercritical water, which show properties in critical-like conditions, are not toxic nor flammable, and are low cost (Agostini et al., 2012). $CO_2$ shows important advantages for extraction by findings based on a literature search comparison with respect to conventional organic solvents as $CO_2$ is completely dissolved in essential oils (Velasco

**TABLE 3.1**
**Critical parameters of different substances potentially useful as supercritical fluids**

| Solvent | Critical temperature (°C) | Critical pressure (Bar) |
|---|---|---|
| Carbon dioxide | 30.95 | 73.76 |
| Water | 373.94 | 220.64 |
| Methane | −82.75 | 46 |
| Ethane | 32.15 | 48.73 |
| Propane | 96.65 | 42.45 |
| Ethylene | 9.25 | 50.35 |
| Propylene | 91.75 | 46 |
| Methanol | 239.45 | 80.85 |
| Ethanol | 240.75 | 61.40 |
| Acetylene | 36 | 62.47 |
| Butane | 152 | 70.6 |
| Ether | 193.6 | 63.8 |
| Pentane | 196 | 32.9 |
| Ammonia | 132.5 | 109.9 |
| Acetone | 234.95 | 47.01 |

*Note:* Taken from Extracción supercrítica de antioxidantes naturales a partir de hierbas y especias (p. 16), by C. Hernández & J. A. Guerrero Beltran, (2009), Temas selectos de ingeniería de alimentos, 3(1).

et al., 2007), amd extracts with no solvent residue are obtained because no energy needs to be applied to evaporate the extractant after extraction since it disappears spontaneously as pressure is reduced (Hernández & Guerrero Beltrán, 2009). By using $CO_2$, a great advantage is gained due to the quality of oil collected by this method being higher. In accordance with the above, a general context of the SFE and the solvents used for this method was provided. From this point on, an insight into $CO_2$ SFE is given, from the equipment used for this solvent to the variables that intervene in the extraction procedure.

### EQUIPMENT

SFE can be carried out by means of equipment that has specific characteristics such as withstanding high pressures (up to 50 MPa). In addition, monitoring of the extraction temperature is required. An example of the equipment, characteristics, and parts is evidenced in Figure 3.1. An SFE apparatus must have:

(a) $CO_2$ tank or cylinder (used to feed the solvent system).
(b) A pressurization pump (pushes and pumps fluid to the extraction column) that maintains the pressure from the pump through the pressure vessel.
(c) Pump controller motor.
(d) Temperature regulator.
(e) Extraction column.

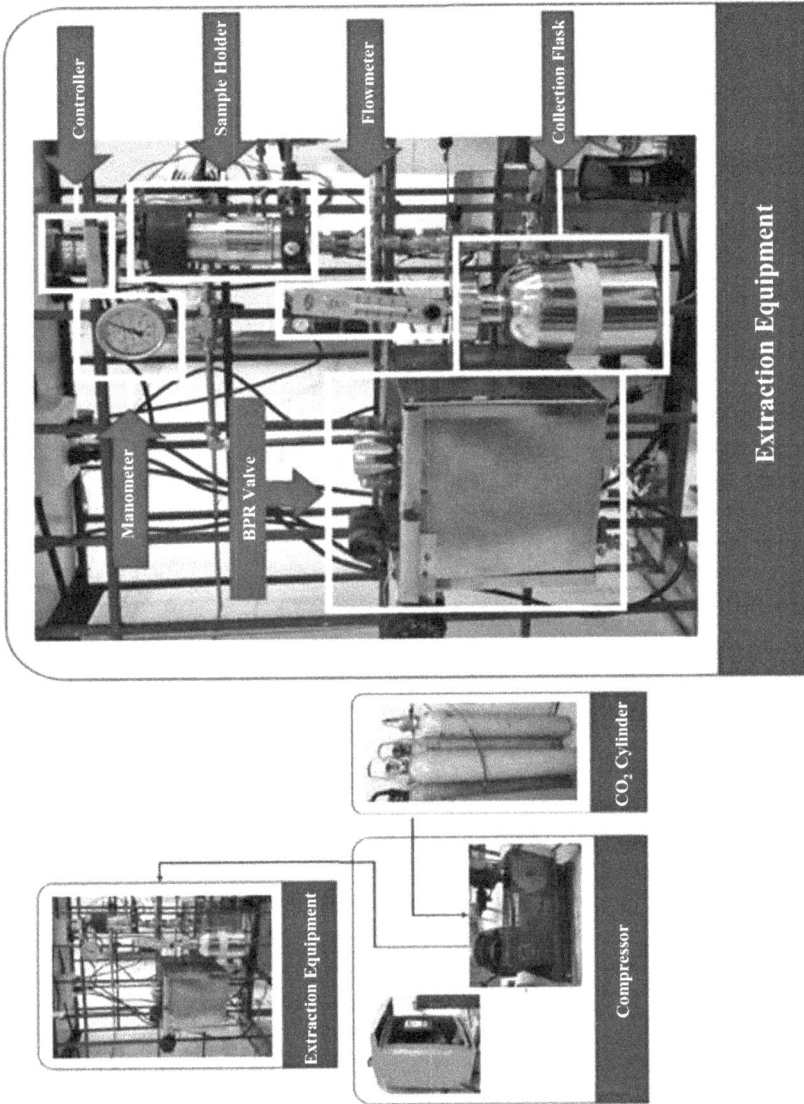

**FIGURE 3.1** General schematics of an extraction equipment and parts.

(f)  Condensation equipment to store $CO_2$ (Cuellar Aquino, 2017; Hernández Romero, 2017).

(g)  Cooling or heating system to achieve the $CO_2$ adiabatic expansion.

(h)  A valve working as separation mechanism between the supercritical fluid and the analyte.

(i)  A BPR valve that allows controlling the $CO_2$ flow rate (an estimated range between 10 to 30 mL/min is recommended) (Cerón et al., 2016; Dorado et al., 2016) which circulates through a ringed outlet tube.

(j)  A container to store the obtained extract (Castro, 2013).

## INFLUENCE OF THE OPERATION VARIABLES

### Density

One of the variables that should be highlighted for supercritical extraction development is the solvent density as this directly relates to pressure and temperature.

If the temperature is constant, the density of a supercritical fluid increases as the pressure increases, which also increases its solubility, causing it to be a very good solvent. Another way to achieve an increase in density is by lowering the temperature when the pressure is constant (Herrera de Pablo, 2002). Maintaining control of $CO_2$ density is important since it facilitates the potential fragmentation of the extracted compounds by means of depressurization as by being above the critical point it changes due to pressure and temperature, more closely resembling the characteristics of liquids than those of gases (Vázquez de frutos, 2008).

### Pressure Influence

Pressure is the most relevant factor in SFE as it is effectively linked to extraction performance. This is because as the pressure is increased, the supercritical carbon dioxide (SC $CO_2$) experiences an increase which is reflected in a higher solvation power to the extraction matrix, allowing for greater solubility by increasing extraction yields (Herrera de Pablo, 2002). On the other hand, keeping a constant high pressure for complex matrices is not suggested in view of the fact that non-required compounds may be extracted, generating a complex analysis and characterization. On the other hand, the presence of coextracted solutes that is achievable and can radically change the level of solubility (Camel et al., 1993) should be taken into account, as it facilitates solubility modifications with slight pressure changes.

As for product separation, this is due to solubility variation when small pressure changes close to the critical point are exerted, by means of precipitation untransformed reaction products can be separated and are reusable. High pressures affect density, triggering the variation of properties both for solvents, as well as for the rate and reaction constant (Sotelo & Ovejero, 2003).

### Temperature Influence

In SFE, as the pressure remains constant and temperature increases, a decreased density is obtained. Therefore, at elevated temperatures and using a volatile solute, solubility and volatility characteristics are changed, where the former decreases and the latter increases. In contrast, for a non-volatile solute, the elevated temperature results in a decrease in the extraction yield (Shilpi et al., 2013). One of the alterations brought about by temperature variations is fluid viscosity (García Fernández, 2001). On the other hand, mass transfer with increasing temperature changes as the extractable compound's vapor pressure increases.

## Extraction time

The extraction time in SFE is very important because it is related to the extraction yield, if the solvent and matrix contact time in the medium is increased, the extraction yield increases. Knowing the extraction time is very important as this goes hand in hand with process costs, the extraction period, and the alteration or loss of heat exhibited by the compounds of interest depending on the exposure phase. Process analysis is carried out considering the overall extraction curves (yield vs extraction time) that give information on the time required for the extraction process and thus, obtain an advantageous and inexpensive process (Shilpi et al., 2013).

## Solvent Flow

Solvent flow is one of the main factors in SFE as it is a direct, proportional influence on the efficiency of extraction. However, very high flows decrease the solvent–sample contact time (Camel et al., 1993; Herrera de Pablo, 2002). In SFE, the use of continuous flow is very common as the constant passage of solvent through the extraction cell favors complete extraction from the matrix because equilibrium is avoided. Similarly, another variable to consider in continuous flow is the flow rate, as an appropriate flow rate should allow a short contact time between the sample and the solvent allowing the solubilization of the analytes of interest, which is an interesting way to reduce chemical reactions and the degradation of the compounds of interest (Castro, 2013). In the same way, flow diffusivity decreases with respect to temperature and as the pressure increases. The values of a solute in a supercritical fluid always above the values corresponding to conventional liquid solvents have also been determined. This occurs because matter transfer is faster as the solute diffusion coefficients are higher (García Fernández, 2001).

## Extraction Yield

Supercritical $CO_2$-collected extracts are obtained using a certain extraction time. For optimizing the extraction time, an extraction curve is plotted to determine the most appropriate time to attain better results. In this sense, and to increase the extraction yield, the particle size must be defined to a range of 0.2–0.5 mm as when this range is used the extraction yield increases because solutes are released through the particles favoring a higher extraction rate (Castro, 2013). Conversely, if the particle size is smaller than 0.2 mm they can cause the system to overpressurize, thus preventing the solvent from freely passing in the extraction process.

## Total Phenol Content (TPC)

For quantification of phenolic compounds, the Folin-Ciocalteu test is implemented where phenolic compounds are determined at a basic pH proceeding to the development of a blue color determined by spectrophotometry. Absorbance is finally measured and the appropriate calibration curve is performed to reveal the concentration of total phenol content.

The reaction mechanism implemented is a redox reaction as this is a method for measuring antioxidant activity where the fundamental constituent of the Folin-Ciocalteau reagent is phosphomolybdotungstic acid which is reduced by the concentration of phenolic groups, leading to the development of an intense blue color, so that the phenol content is measured by this method (Sánchez-Rangel et al., 2013).

The generation of residues from fruit production is an important source of bioactive compounds such as carotenoids, tocopherols, and phenolic compounds of various molecular weights (i.e., low and high) that can offer beneficial effects for human health (Nunes et al., 2016). They are extra nutritional compounds found in reduced portions in plants and food products (D'amario et al., 2018). These compounds have been of considerable use for pharmaceutical and food industries due to their anti-inflammatory, anti-cancer, and anti-mutagenic effects, as well as their association with a low risk of cardiovascular disease (Da Porto et al., 2013).

## Fatty Acids Profile

Fatty acids are biomolecules composed by lipids made up of hydrogen and carbon chains and also commonly known as fats, which have physiological, immune, and structural functions, and are divided into saturated and unsaturated fats (Cabezas Zábala et al., 2016). Fatty acids may be monounsaturated or polyunsaturated and at room temperature can be found in a liquid state. Benefits obtained from this acids are substantial because they help reduce cholesterol levels, relieve discomfort from diseases such as arthritis, and help in the correct functioning of the central nervous system (Cabezas Zábala et al., 2016). Fatty acids are mainly divided into:

- Omega 3 (α-linoleic acid and EPA), as eicosapentaenoic acid (EPA) and docosahexaenoic acid (DHA). EPA is a fatty acid essential in the regulation of brain functions, while DHA is a structured acid part of cell membranes (Aires et al., 2005).
- Omega 6: gamma linoleic acid (GLA) and arachidonic acid (AA) contain omega 6. Arachidonic acid is not recommended as it is inflammatory (Aires et al., 2005).

Therefore, SFE relies on testing of pressure and temperature conditions to determine the effect of extraction factors related with the extraction in addition to response variables (oil yield, distribution, and fatty acid content).

## METHODOLOGY

### SAMPLE PREPARATION

Grape pomace was provided by Casa Grajales in Valle del Cauca. The plant material was subjected to manual cleaning to remove any components other than the pomace in order to obtain a biomass made up of seeds and peel. Clean samples were subjected

**TABLE 3.2**
**Temperature and pressure conditions under**
**which the extracts were obtained**

| Test N° | Temperature (°C) | Pressure (MPa) |
|---------|------------------|----------------|
| 1 | 40 | 20 |
| 2 | 40 | 30 |
| 3 | 60 | 20 |
| 4 | 60 | 30 |
| 5 | 35.9 | 25 |
| 6 | 64.1 | 25 |
| 7 | 50 | 11.9 |
| 8 | 50 | 32.1 |
| 9 | 50 | 25 |
| 10 | 50 | 25 |
| 11 | 50 | 25 |
| 12 | 50 | 25 |

to milling, screening, and finally particle size testing, where particles sized between 0.2 and 0.5 mm were selected and additionally used for SFE.

## OBTAINING EXTRACTS

Extracts were obtained using as process parameters a time of 4 hours, flow of 5.0 kg/h, and as factors to be tested a pressure of 10–30 MPa and temperature of 40–60°C, where a composite rotable central experimental design formed by a 22 factorial design, axial points with $\alpha = 1.414$, and three repetitions at the central point were carried out resulting in 12 extraction points, as shown in Table 3.2, where the extraction yield and total phenol content were tested as response variables.

## STATISTICAL ANALYSIS

Experimental data were analyzed using the Statgraphics Centurion XVI software with a 95% confidence level and the influence of process parameters on the response surface correlating the extraction factors was determined; then the optimized conditions for the extraction process were selected where P-value = 0.211 and $R^2 = 0.86$ were estimated by the software.

## DETERMINATION OF TOTAL PHENOL CONTENT

Total phenol content quantification was carried out using the Folin-Ciocalteu method (Waterhouse Andrew, 2002). Following the above methodology, a calibration curve was performed by means of standards prepared by dissolving gallic acid in 10 mL of ethanol, for subsequent dilution to 100 mL with water. Standards of 50, 100, 250, and 500 mg/L were obtained and subsequently 20 µL of sample, standard, or blank were added followed by 1.58 mL of water and 100 μL of pure Folin reagent (Sigma; Singleton and Rossi, 1965). The reaction time was 8 minutes. Then, 300 μL of 20% (w/v) sodium carbonate were added, leaving to stand for 2 hours. Finally, the reaction was read at an absorbance of 765 nm and the results were reported as mg of gallic acid (GAE)/g extract. Of note, each test was performed in triplicate.

## FATTY ACIDS COMPOSITION

The fatty acid composition of oils was established by means of gas chromatography of fatty acid methyl esters using the Ce-62 AOCs method. A Scion 436 gas chromatography apparatus with a 30-meter long, stainless steel chromatography column, 0.25 micrometers internal diameter, a stationary phase of polyethylene glycol 0.25 micrometers, with an oven temperature 180–230°C, detector temperature 270°C, and injector temperature of 250°C with a boron trifluoride catalyst in methanol was therefore used. For preparation of methyl esters, a fat weight of 0.10 grams was implemented.

## RESULTS

The analyses conducted on the Statgraphics Centurion XVI software produced a quadratic model both for the percentage of yield as well as for TPC where constants A and B corresponded to temperature and pressure, indicating the P-value and $R^2$ as where $R^2$ is greater, fitting of the model to the data is tighter allowing to determine a correct linear adjustment. Regarding the P-value, this refers to the mean absolute error as the residue's average value is determined with the sole purpose of calculating the optimal point, as evidenced in Table 3.3.

Extraction conditions for each test carried out and their respective extraction yields are shown in Table 3.4, where the highest oil yield (6.69%) was obtained at 30 MPa and 60°C, these being the best experimental conditions.

Figure 3.2 shows the response surface analysis for yield estimated by the Statgraphics Centurion XVI software where temperature is evidenced as the variable providing the best percentage of yield since temperature results significantly influence performance. It can be evidenced that at 20 MPa a high rise in the extraction temperature (60°C) leads to a yield reduction, as at low pressures and high temperatures performance decreases, while at 30 MPa the yield increases because this effect is caused by the increase in solute vapor pressure favored by a temperature increase.

## TABLE 3.3
### Extraction variables and their respective polynomial equation

| Variable | Quadratic model | $R^2$ | P-value |
|---|---|---|---|
| % Yield | $4.7175 - 0.230503 * A + 0.563974 * B + 0.716878 * A^2 + 0.7 * A * B + 0.259375 * B^2$ | 0.86 | 0.211 |
| CTF (mg AGE/g extract) | $3.05 + 0.0367159 * A + 0.653082 * B + 0.013749 * A^2 + 1.45 * A * B + 1.37376 * B^2$ | 0.94 | 0.833 |

**TABLE 3.4**
**Oil extraction results from Isabella grape pomace**

| Test | Pressure MPa | Temperature °C | Yield % | TPC mg GAE/g extract |
|---|---|---|---|---|
| 1 | 20 | 40 | 6.24 ± 0.94 | 5.64 |
| 2 | 30 | 40 | 6.41 ± 0.11 | 3.33 |
| 3 | 20 | 60 | 3.72 ± 0.34 | 3.00 |
| 4 | 30 | 60 | 6.69 ± 0.28 | 6.49 |
| 5 | 25 | 35.9 | 5.94± 0.96 | 2.98 |
| 6 | 25 | 64.1 | 6.22± 0.05 | 2.82 |
| 7 | 11.9 | 50 | 4.68± 0.62 | 4.19 |
| 8 | 32.1 | 50 | 5.65± 0.34 | 7.05 |
| 9 | 25 | 50 | 4.69± 0.16 | 3.05 |
| 10 | 25 | 50 | 4.72± 0.17 | 3.05 |
| 11 | 25 | 50 | 4.74± 0.13 | 3.05 |
| 12 | 25 | 50 | 4.72 0.23 | 3.05 |

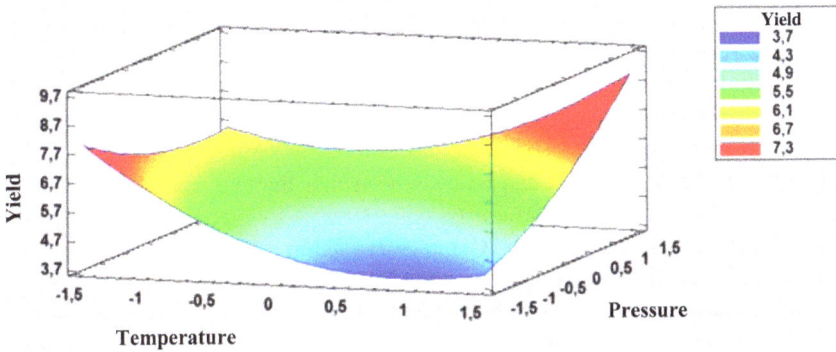

**FIGURE 3.2** Response surface for yield.

Among the results shown in Table 3.4, the concentrations obtained for extract TPC are included and can be found in the range of 2.82–7.05 mg GAE/g extract, where a TPC of 7.05 mg GAE/g extract is reported from experiment 8. On the other hand, a lower TPC (2.82 mg GAE/g extract) was obtained using the conditions of 64.1°C and 25 MPa. This is because at low pressures and high temperatures the oil total phenol content is low as one of the factors that can modify the phenolic profile is the extraction methodology over the inadequate rise in temperature and viticulture factors such as the milling operation, thus causing degradation by oxidation processes. Likewise, the total phenol content increases when there are high temperature (50°C) and pressure (32.1 MPa) values because, for SFE, as these high values are implemented and $CO_2$ is used as solvent, the stability of phenolic compounds in a space scarce of free oxygen is made favorable.

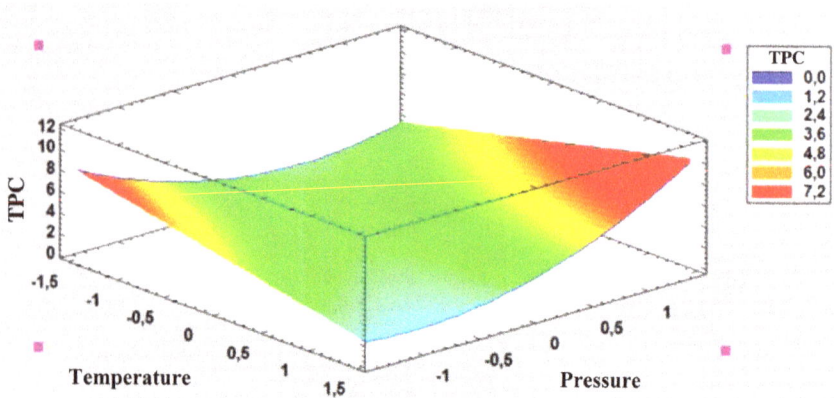

**FIGURE 3.3** Response surface for the TPC variable.

Figure 3.3 shows the pressure and temperature variables with respect to the TPC because as the temperature increased, the TPC amount extracted from Isabella grape oil also increased, as can be observed in this response surface graph. Likewise, as the pressure increased, the TPC yield also increased so that the significantly influencing variable was pressure.

Considering the results of the fatty acid composition specified in Figure 3.4, it is possible to determine that grape seed oil is mainly constituted by unsaturated fatty acids such as linoleic and oleic acids, as well as by saturated fatty acids such as palmitic and stearic acids. It can be evidenced that linoleic acid C18:2 is the one occurring in the highest proportion (63.632 g/100 g) followed by oleic acid C18:1 (17.402 g/100 g), and palmitic acid C16:0 (9.711 g/100 g). Similarly, other fatty acids were found in small amounts. As unsaturated fatty acids showed a high value they are indicative that grape seed oil is a high-quality nutritional oil with benefits for health such as the prevention of some diseases like thrombosis and coronary heart disease, and it also has properties for reducing cholesterol (Sabir et al., 2012).

## DISCUSSION

One of the primary factors in the extraction of grape seed oil is the yield since upon finding the optimal conditions between pressure and temperature the yield tends to improve and, therefore, the amount of oil extracted is increased. In this study different modifications were then introduced to these variables allowing the best yield to be discovered. This study was compared with that of Jokić et al. (2016), where grape seed oil of the grape variety Cabernet Franc. was extracted by SC-$CO_2$ under different extraction conditions, while varying the pressure (20–44 MPa) and temperature (40–64°C). An extraction time of hours and a constant flow rate of 1.94 kg/h were implemented, obtaining an oil yield of 14.81% at 44 MPa and 50°C. Comparing then the SC-$CO_2$ extraction conditions of our study, the pressure used was lower than that proposed by these authors, although the temperature conditions and extraction time were the same.

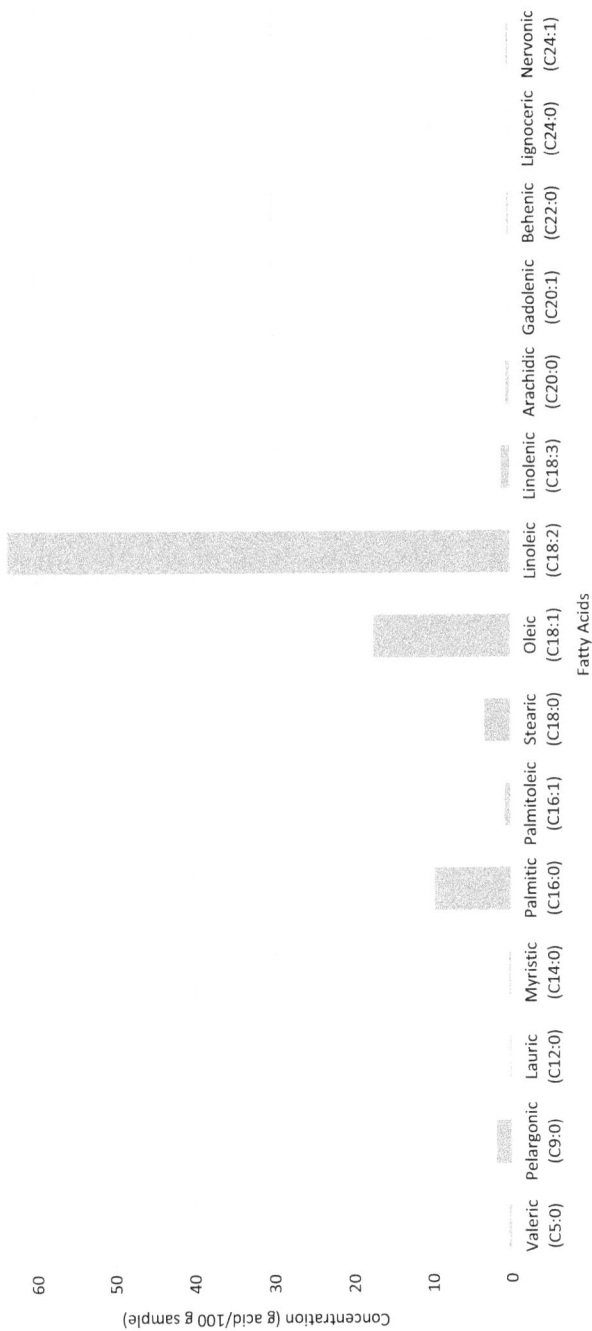

**FIGURE 3.4** Results for fatty acids composition from Isabella grape pomace.

Considering the literature, the extraction yield may be improved at sample preparation. For example, the percentage of moisture in seeds is important since both yield as well as the amount of oil extracted tend to decrease due to moist seeds as Isabella grape seeds tend to be a little woody (Osorio Suárez, 2012). Likewise, establishing adequate grinding is important as by creating a greater contact with the particle surface area, the extraction yield improves because the smaller the particle size is the greater the yield will be (Jokić et al., 2016).

Regarding TPC, it is found in scientific literature that the main components of phenol in oil samples derived from grape peels and seeds are flavonoids and non-flavonoids, where flavanols, flavonols, and anthocyanins are found within flavonoids (Castañeda, 2010). These compounds are valuable because they act as antioxidants that ar more potent than vitamin C as their chemical structure provides protection against oxidative damage. As antioxidants, they prevent rapid aging as well as cholesterol oxidation, avoiding obstruction due to artery clots (Cereceres-Aragón et al., 2019). Taking into account the benefits that TPC brings, research has focused on determining the types of vegetables and fruits that may contain this type of antioxidant since, according to the literature, grape is one of them. In studies, including that by Bañuelos et al. (2017), a comparison was made using the Follin-Ciocalteu method where it a TPC ranging from 4.18 to 22.08 mg GAE/g for the same grape variety as in this study was reported. Compared to our study, where a TPC ranging from 2.82 to 7.05 mg GAE/g using SC- $CO_2$ extraction was obtained, and high TPC values were estimated from both studies, it is then likely that if a high TPC is obtained, antioxidants such as resveratrol will be present. Resveratrol is a natural product derived from grapes with anti-estrogenic activity that inhibits growth of human breast cancer cells. In addition, resveratrol possesses medicinal benefits such as prevention of lung diseases, strengthening of defenses by stimulating the immune system, protection of the body from stress, as well as minimizing heart, inflammatory, allergic, and ulcerative problems (Hidalgo et al., 2016).

On the other hand, the analyzed fatty acid composition showed a large amount of unsaturated fatty acids. Values obtained in this study were compared with several FAMES studies for grape seed oils which are consistent with those reported by other authors (Franco-Mora et al., 2015). For linoleic, oleic, palmitic, and stearic acids with average contents of 71.5, 17.2, 6.6, and 4.3% in wild-type grape (*Vitis* spp.) seed, an 88.7% content of unsaturated fatty acids was also reported. Other studies (Santos et al., 2011) reported the composition of *Vitis labrusca* and *Vitis vinifera* Isabella grape fatty acids in different parts such as the peel, pulp, and seed. The results reported that the main monounsaturated fatty acid (MUFA) in seeds was oleic acid and the largest amount was found in seeds of the Isabela variety, which was 1690.76 mg/100 g. Similarly, in a study presented by Al Juhaimi et al. (2017), in which the properties of grape seed fatty acids were determined in 11 varieties, the results showed that grape seed oils are abundant in oleic, palmitic, and linoleic acids ranging between 13.35–26.30%, 7.15–16.06%, and 47.34–72.91%, respectively. On the other hand, the study presented by Yalcin et al. (2017) examined the composition of fatty acids from different grapes of the *Vitis* genus and showed that linoleic acid and oleic acid were the most abundant, varying between 72.28–72.50% and 13.13–18.50%, respectively. Therefore, it has been proved that grape seed is a good source of vegetable oil and could be used for the extraction of edible oil that exhibits health benefits such as protection from heart disease and the slowing of aging because unsaturated fatty acids such as linoleic acid are an essential element of human cell membranes (Boso et al., 2018). On the other hand, monounsaturated fatty acids such as oleic and palmitic are considered healthy fats as they are acids that do not oxidize or decompose into potentially cancerous substances (Guzmán, 2011).

In addition to the high content of unsaturated fatty acids, other fatty acids were also found in lower concentrations such as stearic acid C18:0, linolenic acid C18:3, and arachidic acid C20:0, which is consistent with those reported by the authors (Boso et al., 2018), where the low concentrations of the lipid profile can vary depending on the oil extraction implemented, since cold or hot extractions can be carried out by different methods. These acids abound both in vegetable and animal fats and are used industrially in the manufacture of candles, soaps, and cosmetics (Guzmán, 2011). Low concentrations of carbon atom odd-numbered fatty acids such as C5:0 valeric acid and C9:0 pelargonic acid were found also; these being the least common because fatty acids naturally contain an even number of carbon atoms as biological synthesis of fatty acids occurs by means of a successive sum of units composed by two carbon atoms. For this reason, fatty acids are usually even (Rodríguez Cruz et al., 2005). Nonetheless, studies have shown that the importance of these fatty acids lies in helping to prevent cardio-metabolic diseases such as hypertension and diabetes, as plasma concentrations of odd-numbered fatty acid chains are related to a lower risk of disease (Jenkins et al., 2015).

## CONCLUSIONS

The extraction of vegetable oil from Isabella grape (*Vitis labrusca*) pomace using SC $CO_2$ was possible where at 30 MPa and 60°C the highest oil yield was obtained (6.69%) taking into account process parameters such as the 4-hour extraction time and a constant flow of $CO_2$ (5.0 kg/h), with temperature being the variable that presented the greatest significance for yield percentage.

For TPC, pressure was identifiable as the most significant variable; however, temperature is also important to obtain good results determining that at high pressure and temperature TPC is higher, obtaining a value of 7.05 mg GAE/g extract under 32.1 MPa and 50°C conditions. The significance of the pressure variable is much lower, giving a not so favorable result of 2.82 mg GAE/g extract under 25 MPa and 64.1°C conditions.

Regarding fatty acids composition, it was possible to show that the main fatty acids found in Isabella grape (*Vitis labrusca*) residue oil extract are linoleic C18:2 and oleic C18:1, with values of 63.632 g/100 g and 17.402 g/100 g, respectively. It is concluded that Isabella grape seeds are a great source of nutritional interest so that the oil extracted from grape seeds could be implemented as an edible oil due to several benefits for health and a high content of antioxidants. Because agro-industrial production currently poses several issues due to the agro-industrial waste, taking advantage of Isabella grape residues would reduce the environmental impacts that agro-industrial businesses generate in water, soil, and air, as they are potentially applicable in different industries such as cosmetics and food.

## GLOSSARY

**Chemical industry:** A series of economic activities focused on obtaining and processing materials or compounds by means of chemical processes.

**Critical point:** Related to equivalence between liquid and vapor density.

**Diffusivity:** A coefficient that provides the ease of yielding a material by means of contact or surface.

**Extractant:** Substance or compound that chemically releases or separates other substances from a more complex mixture.

**Extraction yield:** The molar fraction of the solute in the solvent that behaves as an extractant. Calculated by the relative difference between the unit and molar fraction of the solute retained in the initial solvent (usually water). Subsequently, the yield can be calculated using a general expression and a constant.

**Gallic acid:** A natural phenolic acid that has the advantage of high solubility in water, low price, and wide commercial availability.

**Grape pomace:** The solid part resulting from wine production and composed of grape stalks (25%), seeds (25%), and peels (50%), and considered to potentially generate a negative effect on the environment.

**Isabella grape:** Fruit obtained from the vine. It is a juicy berry that grows in clusters and rounded in shape. The Isabella grape is a shrub whose scientific name is *Vitis labrusca* L. and its cultivation is derived from the *Vitis* grape species.

**Organic solvent:** Chemical compound used to dissolve substances which may contain carbon based on its characteristics.

**Solvent:** Also named dissolvent, this is a substance with dissolution capabilities due to its characteristics to achieve the dilution of a solute of certain chemical composition.

**Supercritical carbon dioxide:** Because of its characteristics, carbon dioxide behaves as a gas or solid depending on the temperature and pressure it is found in. Carbon dioxide behaves as a supercritical fluid under the right conditions.

**Total phenol content:** Determination of total phenol from a sample or plant extract using basic pH.

**Vaporization chamber:** A part of the gas chromatography apparatus formed by a vertical cylinder enclosed by bases with an outlet for the solvent in gas state and other outlet for concentrated solutions.

**Waste oil:** Derived from petroleum as a waste product from the use of oils in a wide range of industrial and commercial activities such as engineering, power generation, and vehicle maintenance. Waste oils must be properly disposed of or treated for reuse.

## REFERENCES

Agostini, F., Bertussi, R. A., Agostini, G., Atti Dos Santos, A. C., Rossato, M., & Vanderlinde, R. (2012). Supercritical extraction from vinification residues: Fatty acids, α-tocopherol, and phenolic compounds in the oil seeds from different varieties of grape. The Scientific World Journal, 2012(May). https://doi.org/10.1100/2012/790486

Al Juhaimi, F., Geçgel, Gülcü, M., Hamurcu, M., & Özcan, M. M. (2017). Bioactive properties, fatty acid composition and mineral contents of grape seed and oils. South African Journal of Enology and Viticulture, 38(1), 103–108. https://doi.org/10.21548/38-1-1042

Angela Viviana Ruales Salcedo, A. F. R. G., & Alzate, C. A. (2017). Obtención de compuestos fenólicos a partir de residuos de uva Isabela (Vitis labrusca) TT – Phenolic compound recovery from isabella grape's waste (Vitis labrusca) TT – obtençåo de compostos fenólicos a partir de resíduos de uva isabella (Vitis labrusca) Biotecnología En El Sector Agropecuario y Agroindustrial, 15(spe2), 72–79.

Aires, D., Capdevila, N., & José Segundo, M. (2005). Ácidos grasos esenciales Offarm. 24, 96–102.

Akanda, M. J. H., Sarker, M. Z. I., Ferdosh, S., Manap, M. Y. A., Rahman, N. N. N. A., & Kadir, M. O. A. (2012). Applications of supercritical fluid extraction (SFE) of palm oil and oil from natural sources. Molecules,17(2),1764–1794. https://doi.org/10.3390/molecules17021764

Bañuelos, F., Martinez, C., Carranza, J., & Concha, J. (2017). Contenido de fenoles totales y capacidad antioxidante de uvas no nativas para vino cultivadas en zacatecas, México.

Bhusnure, O. ., Gholve, S. ., Giram, P. ., Borsure, V. ., Jadhav, P. ., Satpute, V. ., & Sangshetti, J. N. (2016). Importance of Supercritical Fluid Extraction Techniques in. Indo American Journal of Pharmaceutical Research, January 2015.

Brunner, G. (2014). Gas Extraction: An Introduction to Fundamentals of Supercritical Fluids and... – Gerd Brunner – Google Libros (Vol. 4).

Boso, S., Gago, P., Santiago, J. L., Rodríguez-Canas, E., & Martínez, M. C. (2018). New monovarietal grape seed oils derived from white grape bagasse generated on an industrial scale at a winemaking plant. LWT, 92, 388–394. https://doi.org/10.1016/j.lwt.2018.02.055

Cabezas-Zábala, C. C., Hernández-Torres, B. C., & Vargas-Zárate, M. (2016). Fat and oils: Effects on health and global regulation. Revista Facultad de Medicina, 64(4), 761–768. https://doi.org/10.15446/revfacmed.v64n4.53684

Camel, V., Tambuté, A., & Caude, M. (1993). Analytical-scale supercritical fluid extraction: a promising technique for the determination of pollutants in environmental matrices. Journal of Chromatography A (Vol. 642, Issues 1–2, pp. 263–281). Elsevier. https://doi.org/10.1016/0021-9673(93)80093-N

Campalani, C., Amadio, E., Zanini, S., Dall'Acqua, S., Panozzo, M., Ferrari, S., De Nadai, G., Francescato, S., Selva, M., & Perosa, A. (2020). Supercritical CO2 as a green solvent for the circular economy: Extraction of fatty acids from fruit pomace. Journal of $CO_2$ Utilization, 41, 101259. https://doi.org/10.1016/j.jcou.2020.101259

Castañeda, B. (2010). Inducción de antocianinas y capacidad antioxidante por oligogalacturónidos en uvas de mesa cv. 'Flame Seedless.' 84.

Castro, V. H. I. (2013). Obtención de antioxidantes a partir de residuos frutículas empleando extracción con fluidos supercríticos EFS. 164.

Cereceres-Aragón, A., Rodrigo-García, J., Álvarez-Parrilla, E., & Rodríguez-Tadeo, A. (2019). Consumption of phenolic compounds in the elderly population. Nutricion Hospitalaria, 36(2), 470–478. https://doi.org/10.20960/nh.2171

Cerón, L. J., Hurtado, A. M., & Ayala, A. A. (2016). Efecto de la presión y la temperatura de extracción con CO2 supercrítico sobre el rendimiento y composición de xyaba (*Psidium guajava*). Informacion Tecnologica, 27(6), 249–258.https://doi.org/10.4067/S0718-07642016000600025

Cuellar Aquino, R. (2017). Diseño de la automatización para una planta piloto de extracción por fluido supercrítico utilizando $CO_2$ como solvente. [Tesis de Maestrìa,

Pontificia Universidad Católica del Perú]. http://tesis.pucp.edu.pe/repositorio/handle/20.500.12404/8438

D'amario, A. M., Fontana, A., & Antoniolli, A. (2018). Extracción y caracterización de compuestos bioactivos remanentes en orujos y su utilización en la industria alimentaria con fines tecnológicos. [Tesis de Pregrado Universidad Nacional de Cuyo – Facul]. https://bdigital.uncuyo.edu.ar/objetos_digitales/11475/tesis-brom.-damario-2018.pdf

Da Porto, C., Porretto, E., & Decorti, D. (2013). Comparison of ultrasound-assisted extraction with conventional extraction methods of oil and polyphenols from grape (*Vitis vinifera* L.) seeds. Ultrasonics Sonochemistry, 20(4), 1076–1080. https://doi.org/10.1016/j.ultso nch.2012.12.002

Da Silva, R. P. F. F., Rocha-Santos, T. A. P., & Duarte, A. C. (2016). Supercritical fluid extraction of bioactive compounds. TrAC – Trends in Analytical Chemistry, 76, 40–51. https://doi.org/10.1016/j.trac.2015.11.013

Dorado, D. J., Hurtado-Benavides, A. M., & Martínez-Correa, H. A. (2016). Extracción con $CO_2$ Supercrítico de Aceite de Semillas de Guanábana (Annona muricata): Cinética, Perfil de Ácidos Grasos y Esteroles Extraction of Soursop (Annona muricata) Seed Oil by Supercritical $CO_2$: Kinetic, Fatty Acid and Sterol Profiles. Información Tecnológica, 27(5), 37–48. https://doi.org/10.4067/S0718-07642016000500005

Franco-Mora, O., Salomon-Castañ, J., Morales P., A. A., Castañeda-Vildózola, Á., & Rubí-Arriaga, M. (2015). Fatty acids and parameters of quality in the oil of wild grapes (*Vitis* spp.). Scientia Agropecuaria, 271–278. https://doi.org/10.17268/sci.agrop ecu.2015.04.04

García Fernández, I. (2001). Obtención de aceite de orujo meidante extracción con fluidos supercríticos. [Tesis Doctoral Universidad de Castilla- La mancha España]. https://dial net.unirioja.es/servlet/tesis?codigo=71254

Gancedo Alonso, S. (2018). Impactos ambientales derivados de la producción de vino de la D.O.P cangas. Universidad de oviedo.

Guzmán, A. (2011). Perfil lipídico y contenido de ácidos grasos trans en productos ecuatorianos de mayor consumo. http://repositorio.puce.edu.ec/bitstream/handle/22000/3721/T-PUCE-3366.pdf?sequence=1&isAllowed=y

Hernández, C., & Guerrero Beltran, J. A. (2009). Extracción supercrítica de antioxidantes naturales a partir de hierbas y especias. In temas selectos de ingeniería de alimentos (p. 16).

Hernández Romero, A. (2017). Modelización de la extracción de aceites vegetales con CO2 en condiciones supercríticas. [Tesis de Grado Escuela Técnica Superior de Ingeniería Industrial de Barcelona]. https://upcommons.upc.edu/handle/2117/104858

Herrera de Pablo, J. C. (2002). Aplicación de las técnicas de extracción con fluidos supercríticos (SFE). Universidad de Almeria.

Herrero, M., Cifuentes, A., Ibañez, E., da Silva, R. P. F. F., Rocha-Santos, T. A. P., & Duarte, A.C. (2016). Sub- and supercritical fluid extraction of functional ingredients from different natural sources: Plants, food-by-products, algae and microalgae – A review. TrAC – Trends in Analytical Chemistry, 76(1), 40–51. https://doi.org/10.1016/j.trac.2015.11.013

Hidalgo, R., Gómez, M., Soliz, M., Soliz, R., Quiroga, D., González, G., & Saavedra, D. (2016). Propiedades medicinales de la semilla de uva. Revista de Investigación e Información En Salud, 11(N 2075-6194), 54–56.

Jenkins, B., West, J. A., & Koulman, A. (2015). A review of odd-chain fatty acid metabolism and the role of pentadecanoic acid (C15:0) and heptadecanoic acid (C17:0) in health and disease. Molecules, 20(2), 2425–2444. https://doi.org/10.3390/molecules20022425

Jokić, S., Bijuk, M., Aladić, K., Bilić, M., & Molnar, M. (2016). Optimisation of supercritical $CO_2$ extraction of grape seed oil using response surface methodology. In International Journal of Food Science and Technology (Vol. 51, Issue 2, pp. 403–410). Blackwell Publishing Ltd. https://doi.org/10.1111/ijfs.12986

Manjare, S. D., & Dhingra, K. (2019). Supercritical fluids in separation and purification: A review. Materials Science for Energy Technologies, 2(3), 463–484. https://doi.org/10.1016/j.mset.2019.04.005

Nunes, I. L., Mendonça, T. A., Bortolin, R. C., Jablonski, A., Flôres, S. H., De Oliveira Rios, A., & Assumpção, C. F. (2016). Bioactive Compounds and Stability of Organic and Conventional Vitis labrusca Grape Seed Oils. JAOCS, Journal of the American Oil Chemists' Society, 93(1), 115–124. https://doi.org/10.1007/s11746-015-2742-0

Osorio Suárez, L. (2012). Obtención y caracterización del aceite de las semillas de vitis labrusca l. (uva isabella) y evaluación de su actividad antioxidante. Экономика Региона, 32.

Rivera Narváez, Carlos M.; Cardona Bermúdez, Liliana M.; Muñoz, Laura M.; Gómez, Dorely D.; Passaro Carvalho, Catarina.; Quinceno Rico, J. M. (2016). Guía De Extracción Por Fluidos Supercríticos: Fundamentos Y Aplicaciones. 48.

Rodríguez Cruz, M., Tovar, A., del Prado, M., & Torres, N. (2005). Mecanismos moleculares de acción de los ácidos grasos poliinsaturados y sus beneficios en la salud. Revista de Investigación Clínica, 57(03). www.scielo.org.mx/scielo.php?script=sci_arttext&pid=S0034-83762005000300010

Sánchez-Rangel, J. C., Benavides, J., Heredia, J. B., Cisneros-Zevallos, L., & Jacobo-Velázquez, D. A. (2013). The Folin-Ciocalteu assay revisited: Improvement of its specificity for total phenolic content determination. Analytical Methods, 5(21), 5990–5999. https://doi.org/10.1039/c3ay41125g

Sabir, A., Unver, A., & Kara, Z. (2012). The fatty acid and tocopherol constituents of the seed oil extracted from 21 grape varieties (*Vitis* spp.). Journal of the Science of Food and Agriculture, 92(9), 1982–1987. https://doi.org/10.1002/jsfa.5571

Santos, L. P., Morais, D. R., Souza, N. E., Cottica, S. M., Boroski, M., & Visentainer, J. V. (2011). Phenolic compounds and fatty acids in different parts of *Vitis labrusca* and *V. vinifera* grapes. Food Research International,44(5),1414–1418. https://doi.org/10.1016/j.foodres.2011.02.022

Shilpi, A., Shivhare, U. S., & Basu, S. (2013). Supercritical CO2 Extraction of Compounds with Antioxidant Activity from Fruits and Vegetables Waste-A Review. Focusing on Modern Food Industry (FMFI), 2(1), 43–62. www.fmfi-journal.org

Sotelo Sancho, J., & Ovejero Escudero, G. (2003). Procesos con fluidos supercríticos. Anales de La Real Sociedad Española de Química, 4, 15–23.

Sovovà, H., & Stateva, R. P. (2011). Supercritical fluid extraction from vegetable materials. Reviews in Chemical Engineering, 27(3–4), 79–156. https://doi.org/10.1515/REVCE.2011.002

Vardanega, R., Prado, J. M., & Meireles, M. A. A. (2015). Adding value to agri-food residues by means of supercritical technology. Journal of Supercritical Fluids, 96, 217–227. https://doi.org/10.1016/j.supflu.2014.09.029

Vargas Reyes, J. (2018). Extracción con fluidos supercríticos: aplicaciones de interés farmacéutico. [TesisdegradoUniversidad dedegra. https://idus.us.es/handle/11441/82249

Vázquez de frutos, L. (2008). Extracción con fluidos supercríticos y síntesis enzimática para la obtención de lípidos funcionales luis vázquez de frutos. [Tesis Doctoral Universidad AutónomadeMadrid]. https://repositorio.uam.es/bitstream/handle/10486/1830/5441_vazquez_frutos_luis.pdf?seq uence=1&isAllowed=y

Velasco, R. J., Villada, H. S., & Carrera, J. E. (2007). Aplicaciones de los Fluidos Supercríticos en la Agroindustria. Applications of supercritical fluids in the agroindustry. Información Tecnológica, 18(1), 53–65. https://scielo.conicyt.cl/pdf/infotec/v18n1/art09.pdf

Waterhouse Andrew.L. (2002). Determination of total phenolics. Current Protocols in Food Analytical Chemistry, 1–8.

Yalcin, H., Kavuncuoglu, H., Ekici, L., & Sagdic, O. (2017). Determination of fatty acid composition, volatile components, physico-chemical and bioactive properties of grape (*Vitis vinifera*) seed and seed oil. Journal of Food Processing and Preservation, 41(2). https://doi.org/10.1111/jfpp.12854

# 4 Recovery of Phenolic Antioxidants from Isabella Grape (*Vitis labrusca* L.) Pomace Using Supercritical Carbon Dioxide with Added Ethanol as Co-Solvent

*Jenny Viviana Bejarano Pérez, Jessica Tatiana Mancera Cifuentes, Patricia Joyce Pamela Zorro Mateus, and Henry Isaac Castro Vargas*

In the wine-making industry during grape processing by-products called agro-industrial waste are generated which are mainly composed of seeds, stalks, peels, and leaves (grape pomace). These by-products represent about 18% of the grape (Soler Fernández, 2017) and they generate final waste disposal issues within the economic and environmental settings. Valorization of the by-products is therefore sought since benefits more substantial than those from grape's own pulp may result due to bioactive compounds such as polyphenols, resveratrol, anthocyanins, proanthocyanidins, and quercetin within the by-products and oil in the seed. For polyphenols, antioxidant characteristics and a significant health benefit (Casas et al., 2008) have been found as polyphenols lower the risk of contracting carcinogenic diseases, stop the proliferation of malignant cells, and act as anti-inflammatory substances (Ramírez, 2019). Polyphenols also present themselves as natural additives that can be used in the food industry.

Recovery of these bioactive compounds may be carried out by using several extraction methods. Green, environmentally friendly technologies like carbon dioxide added, supercritical fluid extraction (SFE) with ethanol as co-solvent ($CO_2$-EtOH) are currently chosen, where their function is to separate the extractant from the matrix using a supercritical fluid that functions as an extraction solvent (Sapkale, 2010). This technology has proven to be technically and economically viable over several other traditional methods using organic solvents (Cuellar, 2017) because changes in density are permitted, leading to variations in temperature and pressure. In addition, no residual chemicals are left in the process and $CO_2$ extraction favors recycling of $CO_2$ to be used again. The most commonly used supercritical fluid is $CO_2$ as it is non-toxic, odorless, inert, non-flammable, and inexpensive. A modifier or co-solvent can be added to $CO_2$ in order to speed up the extraction process and generate modified mixtures where extraction of natural, polar target solutes is possible (Palacios et al., 2013). Among the modifiers, EtOH is found to deliver a greater extraction efficiency and selectivity because it is polar, providing better results in yield and for obtaining bioactive compounds.

The purpose of this research was to test a method to optimize the residues generated by the Casa Grajales wine-making industry through implementation of green technologies such as supercritical fluid extraction with $CO_2$-EtOH as a co-solvent so that the Isabella grape (*Vitis labrusca* L.) pomace phenolic and antioxidant compounds are extracted. This study was developed in phases. Initially, raw matter (Isabella grape pomace) was collected and the respective cleaning, grinding, and sieving were carried out. Supercritical fluid extractions using $CO_2$ and 5%, 10%, and 15% EtOH were then performed and the extract collected was further analyzed to obtain the total phenol, flavonoids, and antioxidants content. Finally, the extract was tested as an additive for an edible oil, to measure the induction time by means of a Rancimat apparatus.

## THEORETICAL FRAMEWORK

The agro-industry is considered to involve the transformation of products from agriculture, forestry, and fishing activities and consists of a series of processes for producing the raw materials derived from the agricultural sector as a supply source in the manufacture of other goods. A significant portion of agricultural production undergoes a certain degree of processing from harvesting through to the final use, generating residues such as seeds, stems, leaves, and peels, among other by-products described as agro-industrial residues. Such waste has become a global problem due to inadequate management and final disposal, constituting potential sources of pollution and health-related risks (Guerrero et al., 2011). As shown by Barragán et al. in 2008, at waste management some residues are incinerated or taken directly to the sanitary landfill, releasing a large amount of carbon dioxide, with contamination of water sources, adverse effects of bad odors, and pest proliferation, among other negative effects. Contamination may also occur mainly in soil and water resources, both in surface and underground sources (Guerrero and Valenzuela, 2011). Emissions from agro-industrial operations which may include particulate matter, sulfur oxide, nitrous oxides, hydrocarbons, and other organic compounds result in atmospheric contamination to a lesser extent.

Considering the above, some properties and potential uses are attributed to agro-industrial waste as they are a source of potentially bioactive compounds like additives or supplements (food, cosmetic, and pharmaceutical) with a high added and commercial value. However, in Colombia, information about the properties of agro-industrial by-products and recovery methods for the effective use of these waste is limited.

One of the world's most important industries is related to the culture and profit from grapes mainly used as a raw material in the production of fermented beverages such as wine. In the wine-making industry, by-products from the Isabella grape (*Vitis labrusca*. L) are composed of seeds, peels, and brooms (grape pomace) which can be used for obtaining bioactive compounds, for example, phenolic compounds (phytophenols) and antioxidants (flavones) where most bioactive compounds are in the peel, pulp, and seeds (Barros et al., 2014). Bioactive compounds help to inhibit platelet aggregation, play a role as heart-protective agents, and function as anti-inflammatory compounds and anti-cancer agents.

Studies have shown that grape extracts are active against HIV by inhibiting virus expression and replication (Salcedo et al., 2017). On the other hand, these phenolic and antioxidant compounds are important for human health as their effects are of cardiovascular benefit, increase antioxidant capacity and resistance to low-density lipoproteins, improve endothelial function, and reduce the risk of chronic diseases, hypertension, and cancer (Barros et al., 2014; Rahman et al., 2022). Phenolic compounds and antioxidants can also be used in different therapeutic procedures intended to neutralize the 2,2'-diphenyl-1-picrylhydrazyl free radical in biological systems (Ghafoor et al., 2010; Michalak, 2022). Recently, grape and wine phenols, particularly those obtained from winery by-products, have also drawn attention due to their potential application as food antioxidants (González et al., 2004; Pinelo et al., 2006) and for their protective factor as natural additives.

According to the properties and benefits of the target residue, compounds with antioxidant activity that might be used in the formulation of items including, but not limited to, moisturizing creams, makeup, functional foods, enriched foods, lubricant stabilizers, and fuel antioxidants are obtainable and further uses for such compounds are deliverable due to the recovery of a variety of products from pomace such as ethanol, tartrates, citric acid, grape seed oil, hydrocolloids, and dietary fiber (Segura et al., 2015). As it is the case of grape oil which has excellent properties for direct consumption due to the high content of essential fatty acids and natural antioxidants, this oil can be used as a frying medium due to its high smoking point index or used topically in skin treatments, cosmetics, and aesthetic products generally (Martínez and Ceballos, 2012).

Currently, a need to incorporate technologies that allow the recovery and valorization of by-products generated by the wine-making agri-business has risen due to their natural properties that offer health benefits and that they are produced by bioactive compounds including polyphenols, mainly located in peels and seeds depending on the variety, climate, soil for cropping, and techniques used for harvesting. Polyphenols are synthesized by plants as they are essential for plant physiology and defense. Among these compounds the flavonoids, the largest polyphenol subgroup and their main role as a protector against oxidizing agents can be highlighted. The increased molecular content and antioxidant capacity of flavonoids are related to the intensity of light to which the fruit is exposed. In this sense, conventional methods for extracting compounds of interest are employed in the agro-industry which during the process use large volumes of toxic solvents, prolonged operation times, and elevated working temperatures that accelerate degradation of the expected products, modifying their structure or allowing the formation of other metabolites, in addition to being harmful to the environment and human health (Velásquez, 2008). The development and application of green extraction techniques which are amicable with the environment as they decrease the amount of solvent, require less time and extraction energy, and allow the use of green solvents such as ethanol, water, or their mixtures have been revealed in recent studies (Castejón et al., 2018).

An increasingly used green extraction technique is supercritical fluid extraction (SFE), which is characterized by its properties to extract compounds of interest using certain solvents under the combination of high pressures and low temperatures (Velasco et al., 2007; Da Porto et al., 2014). A solvent is required to carry out the SFE and, among the existing solvents, carbon dioxide ($CO_2$) is the most commonly used given its properties such as low cost, and no toxicity, flammability, or corrosiveness; it is easily removed and leaves no residues due to the low viscosity and high diffusivity that allow improved extraction speed and efficacy (Luque de Castro et al., 1993).

The $CO_2$ non-polar characteristics define it as a weak solvent for the extraction of high-polarity substances such as some phenols with antioxidant effect and proteins in grape pomace (Zulkafli et al., 2014). $CO_2$ weakness as a solvent is the reason why the addition of a polar modifier (co-solvent) has been evaluated in various studies because a polar modifier increases the process effectiveness and efficiency because of the increased solubility caused by the solute and co-solvent physical–chemical interaction (Santos et al., 2017). Likewise, polar modifiers cause a swelling in the matrix that allows solute transportation. Among the mostly used modifiers or co-solvents, ethanol (EtOH) is found where different polar and ionic solutes such as proteins, carbohydrates, and mineral salts are soluble. The co-solvent is used for extracting bioactive compounds due to its low toxicity. This modifier exerts an influence on the procedure selectivity which, once a mixture with $CO_2$ takes place, changes the polarity of the flowing extract and improves the solvation capability (Gavilan, 2016) allowing the extraction time to be reduced. According to studies carried out by Vatai in 2009, twice as much phenol can be obtained using $CO_2$ in supercritical fluid extraction (SFE-$CO_2$) adding EtOH as co-solvent in extractions from grape pomace compared with SFE using pure $CO_2$ (Vatai, 2009). Due to its characteristics, the apparatus is washed using EtOH at the end of the extraction phase thus presenting advantages vs. equipment clean up, as compared to $CO_2$ extractions carried out with no modifier.

In a study conducted by Herrera and collaborators in 2007, extractions where EtOH was used as co-solvent resulted in a predominant extraction of total phenolic compounds, especially recovering total flavonoids. It is noteworthy that the percent of modifier obtaining the highest yield during extraction was 15% EtOH (Herrera et al., 2007). Several studies have shown that EtOH works in small quantities within the 5–15% range (Farías-Campomanes, 2013) where the best results in the process are obtained. Research developed by Farías et al. in 2013 shows that the content of Soxhlet-extracted (conventional method) phenol was 1.8 g/kg extract compared to 10% EtOH-added SFE at 20 MPa whose yield was 23.0 g/kg extract, demonstrating more significant extraction of phenolic compounds (Farías et al., 2013).

Regarding the analysis of antioxidant activity in grape residue extracts the use of the Rancimat apparatus renders itself a solution because determination of the extract induction periods by means of oxidative stability of different oils or vegetable fats is allowed. As shown by the work carried out by Yalcin et al. (2017), where the aim was to ascertain the effect on corn oil oxidative stability using five types of grape seeds (Kalecik Karasi, Cabernet, Gamay, Okuzgozu, and Senso) testing was subjected to conditions of 120°C and a flow rate of 20 L/h using 3 g of oil and 2,000 ppm of extract. The results showed a protection index of 3.18 h for the control, being lower than extract containing samples whose results ranged between 3.31 and 3.41 h, and Gamay grape seed extracts were the most effective (Yalcin et al., 2017). This study shows that grape extracts have a protective factor that may serve as a food natural antioxidant as synthetic antioxidants are mainly used in this industry due to their low cost and effectiveness that reduce emerging oxidative phenomena, avoid food rancidity, and extending the shelf-life of the product. Butyl-hydroxy-anisole or E-320 (BHA), butyl-hydroxy-toluene or E-321 (BHT), propyl gallate (PG), and tert-butyl-hydroquinone (TBHQ) are some synthetic antioxidants; however, they pose some disadvantages as they are quite volatile as well as easily decomposing at high temperatures so their use is limited to certain quantities (Ahn et al., 2007).

If consumed at high doses, some of the antioxidants mentioned above may be detrimental to health. For this reason, the use of some antioxidants in food has been banned in Europe, and in the United States the use of these antioxidants in products has been regulated. For handling BHT, BHA, and TBHQ, limits for authorized concentration levels are provided so that the quality of food is not affected, such as in vegetable oils, where a concentration of 200 mg/kg must not be exceeded whether it is alone or combined (Ding and Zou, 2012), for commercial purposes or laboratory practices, if utilization for human consumption as safety standard is desired.

One of the problems with adopting antioxidant additives from food arises in health, as with TBHQ, hyperplasia and low hemoglobin levels occur (Almeida-Doria and Regitano-D'arce, 2000). The fact that TBHQ is the best synthetic antioxidant does not make it the most suitable for health. According to studies, TBHQ is harmful in laboratory animals, and there is evidence of toxicity and mutations at the cell level ranging from the formation of cancer cells to mutations in bacterial DNA, so a risk of toxicity upon long-term consumption of products containing TBHQ as an additive should be presumed (Espinosa, 2017; Khezerlou et al., 2022).

Therefore, the implementation of natural additives from fruits, leaves, or vegetables in the form of pure compounds, extracts, and/or oils as alternatives for the food industry is sought. For the meat industry, according to a review conducted by Pateiro et al. in 2018, the application of essential oil as a potential substitute of synthetic antioxidants for meat and meat products was evidenced in several studies, where application of one or a combination of two essential oils to protect oxidation is feasible. Essential oils are beneficial to the health of their consumers (Pateiro et al., 2018) and among the advantages of essential extracts and oils are the ability of removal of free radicals, with the antioxidant capacity due to the content of phenol, polyphenol, and flavonoids. Essential extracts and oils bring positive effects to health as the formation of artery plaques and carcinogenic and cardiac diseases are prevented (Anbudhasan et al., 2014).

## METHODOLOGY

### Sample Preparation

The implemented biomass was the Isabella grape (*Vitis labrusca* L.) pomace provided by Casa Grajales company that underwent a separation process where seeds and peel were selected. The sample was dried at room temperature and under shade, and the milling process was then carried out by means of a household, manual mill (Corona brand). The sample was subsequently sieved using a laboratory screen (Fieldmaster brand) where a particle matter size of 420 μm was selected.

### Supercritical Fluid Extraction with $CO_2$-EtOH

The extraction method was performed in triplicate using 5%, 10%, and 15% ethanol with EtOH flow rates of 0.3, 0.6, and 0.9 mL/min, respectively. The $CO_2$ flow control of 5 L/min was regulated by a valve, and temperature and pressure conditions of 60°C and 31.7 MPa, respectively, were kept constant during the process. The extraction time using $CO_2$-EtOH was planned as 5 hours to obtain most of the phenols during that period. Once the extraction process was completed, the co-solvent was separated from the sample using a rotary evaporator. Extract yield was determined by the ratio between the extract total mass and sample initial mass on a wet basis.

### Total Phenol Content

Determination of the total content of phenolic compounds (TPC) was carried out following the microscale protocol for Folin-Ciocalteau colorimetry proposed by Waterhouse (2002). Briefly, 20 μL of extract were taken and 1580 μL of distilled water were added followed by 100 μL of pure Folin-Ciocalteau reagent. The sample was mixed well and incubated at room temperature for 7 minutes. Subsequently, 300 μL of a 20% (w/v) sodium carbonate solution were added and incubated in darkness for 2 hours at room temperature. Finally, sample absorbance was measured at 765 nm. For calculation of the TPC, a gallic acid calibration standard with concentrations of 50, 100, 250, and 500 mg/L was used and the results expressed in gallic acid-equivalent milligrams per gram of extract (mg GAE/g extract) and gallic acid-equivalent milligrams per 100 g of sample (mg GAE/100 g sample).

### Total Flavonoid Content

Following the methodology of Chang et al. (2002), the extract solution was mixed with 40 μL of $AlCl_3$, 40 μL of 1 M sodium acetate, 600 μL of 95% EtOH, and 1120 μL of distilled water; the mixture was stirred and incubated at room temperature for 30 minutes. Absorbance was measured at 415 nm using quercetin (QE) concentrations of 10, 20, 40, 60, 80, and 100 μg/mL. Data were expressed as mg QE/100 g sample.

### Antioxidant Activity

The antioxidant activity of grape pomace extract was tested by DPPH (Magalhães et al., 2012; Tournour et al., 2015) consisting of the ability to remove the

2,2-diphenyl-1-picrylhydrazyl free radical. In the DPPH microplate methodology, 150 μL of the grape pomace extract or Trolox diluted standard solution were used together with 150 μL of ethanolic 50% (v/v) DPPH solution; the microplate was incubated for 60 minutes and then absorbance measured at 570 nm. Antioxidant activity was determined through a calibration curve over an absorbance range from 0 to 50 μM, and was expressed as mg Trolox/100 g sample.

## OIL OXIDATIVE STABILITY (OOS) USING RANCIMAT

Oxidative stability was carried out by the Rancimat method. Antioxidant free soy-bean oil provided by the Duquesa SA company spiked with grape pomace extracts was used as the working material. For testing, extracts obtained at 5%, 10%, and 15% EtOH were used and treated in triplicate, each in the presence of a control. The extracts were then tested at a concentration of 500 ppm, nitrogen was added through removal of moisture content, 5 g soybean oil was added, and mixed using a stirrer for further automated analysis. In addition, oxidative stability of collected extracts was compared against the synthetic antioxidant TBHQ at the same concentration of 500 ppm. The apparatus temperature was set at 120°C with an air flow of 20 L/h (Yalcin, 2017). The results were expressed in hours.

## RESULTS AND DISCUSSION

In this research, SFE $CO_2$-EtOH was used as the extraction method for Isabella grape residues, little literature was found on the addition of a modifier in the extraction process. Furthermore, no studies on Isabella grape pomace from the Colombian agri-business have been carried out, therefore making this project unique compared to other ongoing studies. This section describes the different techniques performed and constitutes a proof of the utility of Isabella grape pomace in agri-business.

### EXTRACTION YIELD

The extraction method was carried out by supercritical fluid $CO_2$-EtOH, making use of the optimal process variables (temperature and pressure) that were obtained by means of the extraction kinetics with time intervals that reached 9.18 hours with a total of 28 sampling points. After the extraction, the yield was analyzed using the extract weight in relation to the pomace sample weight, obtaining the results shown in Table 4.1.

The ratio of the extract yield to the amount of co-solvent applied was linear, with a coefficient of determination value of 0.9929, as illustrated in Equation 4.1.

$$y_1 = 0.3 x_1 + 2.6033 \qquad (4.1)$$

where $y_1$ represents the yield of grape pomace extract (%) and $x_1$ is the percentage of EtOH in the obtained correlation. The behavior of increasing yield relative to EtOH concentration is shown in Figure 4.1.

**TABLE 4.1**
**Extraction yield**

|        | Yield        |
|--------|--------------|
| % EtOH | %Y           |
| 5      | 4.03±0.15    |
| 10     | 5.75±0.08    |
| 15     | 7.03±0.27    |

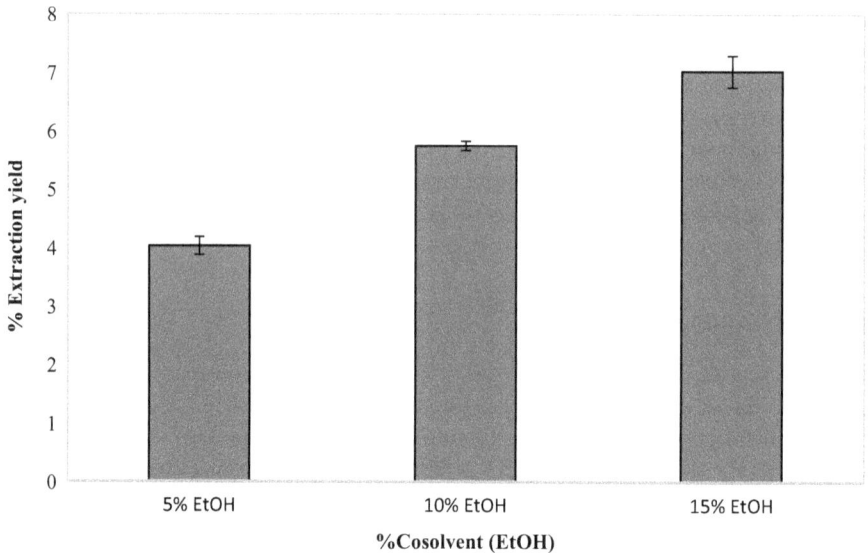

**FIGURE 4.1**  Percentage of extraction yield based on concentrations of 5%, 10%, and 15% (v/v) EtOH.

*Note:* Error bars represent standard deviation bars.

Results obtained using different concentrations of EtOH at 5%, 10%, and 15% (v/v) indicate a higher yield obtained with the 15% EtOH (7.03% w/w) concentration, thus favoring this co-solvent concentration due to proportional changes in the solvent mixture characteristics. Studies such as that by De Campos et al. (2008) show that the best yield results are obtained using 15% ethanol in SFE from grape pomace (Cabernet sauvignon) with a value of 9.2% compared to no co-solvent or application of 10% ethanol where the data range from 2% to 4%. In addition, 15% ethanol has been shown to be the optimal value to achieve the best products as at 20% EtOH the extraction yield is decreased with a value of 6.3% (De Campos et al., 2008). The study conducted by Calvo et al in 2017 on grape seeds demonstrated the same behavior where SFE-$CO_2$ + EtOH as modifier generated a change, and the addition of co-solvent led to a significant increase in the extraction yield of up to 8% by using 15% EtOH and a pressure of 30 MPa. The increased extraction yield could not be achieved at 45 MPa, yielding 4.4% with no modifier (Calvo et al., 2017), which shows evidence that the EtOH content is a substantial variable for obtaining a higher yield. Moreover, the value of the yield depends on the type of grape with which extraction is performed, as shown by Otero-Pareja et al. (2015), where different behaviors were observed only by using 20% EtOH in SFE for six grape pomace varieties under the same conditions and yield ranged from 3% to 7%. The Petit Verdot variety showed the best yield, as opposed to the Tintilla variety (Otero-Pareja et al., 2015), due to the way they are harvested and because they have several bioactive components that make them unique. However, in order to obtain a more significant result vs. yield, 20% EtOH is not indicated as the best value since, according to the literature, the optimal range for wine-making residues lies between 5% and 15% EtOH. Therefore, the EtOH concentrations above were chosen for project development.

## TOTAL PHENOL AND FLAVONOID CONTENT RATIO

By implementing the Folin-Ciocalteau colorimetry method the total phenol content was obtained at the co-solvent study concentrations, as shown in Table 4.2.

The ratio of TPC to the amount of co-solvent used was linear, with a coefficient of determination value of 0.9918, as shown in Equation 4.2.

$$y_2 = 45.805 x_2 + 41.73 \qquad (4.2)$$

where $y_2$ represents the grape pomace extract TPC (mg GAE/100 g sample) and $x_2$ is the EtOH percentage in the correlation obtained. The behavior of increasing TPC relative to EtOH concentration is shown in Figure 4.2.

By means of the aluminum chloride colorimetric method, the total flavonoid content was obtained at the established EtOH concentrations, which are shown in Table 4.3.

### TABLE 4.2
### Total phenol content

| | TPC | |
|---|---|---|
| % EtOH | mg GAE/g extract | mg GAE/100 g sample |
| 5 | 70.11±3.31 | 282.78±13.34 |
| 10 | 82.78±5.21 | 475.73±29.92 |
| 15 | 105.35±5.25 | 740.83±36.91 |

### TABLE 4.3
### Total flavonoid content

| % EtOH | mg QE/g extract | mg QE/100 g sample |
|---|---|---|
| 5 | 47.3 ± 1.036 | 191.0 ± 4.180 |
| 10 | 62.8 ± 1.731 | 360.8 ± 9.950 |
| 15 | 82.1 ± 3.777 | 577.5 ± 26.558 |

The ratio of TFC to the amount of co-solvent employed was linear, with a coefficient of determination value of 0.9953, as shown in Equation 4.3.

$$y_3 = 0.0386\,x_3 - 0.0097 \tag{4.3}$$

where $x_3$ is the TFC (mg QE/100 g sample) and $y_3$ is the percentage of co-solvent. Evidence that at higher EtOH concentrations the content of flavonoids in the extract increases is shown by this equation, thus demonstrating that the extraction method adding co-solvent allows phenolic compounds to be obtained at relatively low conditions. The behavior of increasing TPC and TFC relative to EtOH concentration is shown in Figure 4.2.

The TPC and TFC were expressed as milligrams of extract per 100 grams of sample. Figure 4.5 shows that both TPC and TFC increase as modifier is added. A higher TPC and lower TFC were evidenced, which lies within the expected results as flavonoids belong to a sub-group of phenolic compounds. These results indicate that the extracts obtained possess an antioxidant activity that will be useful for oil rancidity testing as a natural additive. These effects were evidenced in two studies by Ghafoor et al. (2010–2012), where the results of bioactive compounds from Campbell Early grape peels using SFE-CO$_2$ (13.7–16.7 MPa, 37–46°C, and 5–8% EtOH) were significantly affected by temperature, pressure, and to a lesser extent EtOH percentage. However, if lower temperature conditions (37°C) and a higher pressure (16.7 MPa) and co-solvent % (8%) are used for working a TPC of 1.347 mg GAE/100 mL was obtained, unlike if lower pressure conditions of 13.7 MPa and 5% EtOH were used, a set up that resulted in a TPC of 0.887 mg GAE/100 mL (Ghafoor et al., 2010). For the second study, the focus was set on grape (*Vitis labrusca*) seed and the same behavior was found for the modifier when low conditions were chosen; using 8% EtOH a TPC of $1.35 \pm 0.02$ mg GAE/mL extract was obtained compared to using 5% EtOH that led to a TPC result of $0.77 \pm 0.08$ mg/mL extract (Ghafoor et al., 2012).

To demonstrate the phenolic potential of collected extracts, a comparison vs. results in Chapter 3 (Table 3.4) was established where a supercritical fluid extraction

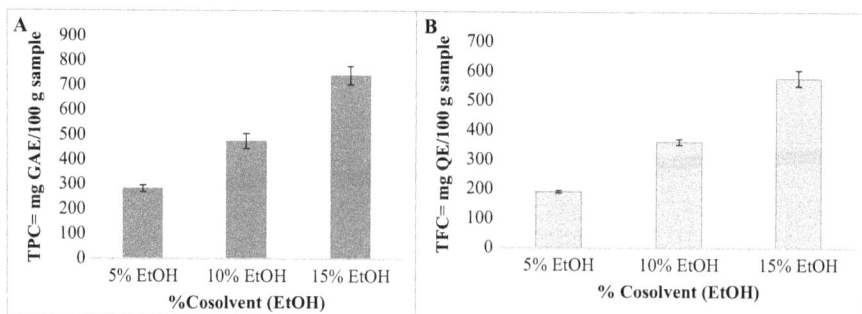

**FIGURE 4.2**   Ratio of the total phenol (A) and flavonoid contents (B) based on 5%, 10%, and 15% (v/v) EtOH concentrations.

*Note:* Error bars represent standard deviation bars.

with $CO_2$ and no modifier and temperature or pressure variations were used. The conditions resembling those of this study were 60°C for temperature and a pressure of 30 MPa resulting in a TPC of 43.40 mg GAE/100 g sample. The data above demonstrate improved results by using a co-solvent in the extraction method since at 5% EtOH the TPC was 282.78 mg GAE/100 g sample and 740.83 mg GAE/100 g at 15% EtOH.

For TFC, the addition of a modifier at 15% resulting in a higher flavonoid content was described by Buelvas in 2018, and according to results from that study on valorization of mango seed, the highest values were displayed at 11 MPa/60°C/15% EtOH (13.63 mg-eq Querc/g extract). It should be noted that TFC extraction with $CO_2$-EtOH was higher compared to extraction by the Soxhlet conventional method, obtaining values up to 105.18 mg-eq Quercetin per 100 g of dry sample (Buelvas, 2018).

## ANTIOXIDANT ACTIVITY

The DPPH methodology followed for grape pomace extracts yielded favorable results as the amount of EtOH increased, as shown in Table 4.4.

The ratio of antioxidant activity to the amount of co-solvent used was linear, with a coefficient of determination value of 0.9879, as illustrated in Equation 4.4.

$$y_4 = 24.192\, x_4 - 48.84 \qquad (4.4)$$

Antioxidant activity is represented by $y_4$ (mg Trolox/100 g extract) and EtOH concentrations of 5%, 10%, and 15% by $x_4$, showing the same trend as former techniques where the amount of ethanol improved the results by increasing the co-solvent content, as shown in Figure 4.3.

The antioxidant activity of extracts was expressed as mg Trolox per 100 g sample and per 100 g extract. The best value was obtained under the following conditions: 37.1 MPa, 60°C, 15% EtOH (321.76 mg Trolox/100 g sample). As evidenced by the TPC and TFC, if the co-solvent percent is increased a considerably increased extract antioxidant capacity is seen due to the DPPH inhibitory properties, corroborating the results obtained in previous phases. Using the same method and ethanol content, Da Porto et al. (2014) showed that the antioxidant activity of grape pomace tested by free radical total elimination capacity had a better value of 698.6 mg tocopherol/100 g of dry matter through studies that compared SFE $CO_2$ + 15% EtOH and methanol extraction, unlike the technique using methanol with a value of 677.9 mg tocopherol/

**TABLE 4.4**
**Antioxidant activity by DPPH method**

DPPH

| % EtOH | mg Trolox/100 g sample | mg Trolox/100 g extract |
|---|---|---|
| 5 | 79.84 | 19.79 |
| 10 | 177.64 | 30.91 |
| 15 | 321.76 | 45.75 |

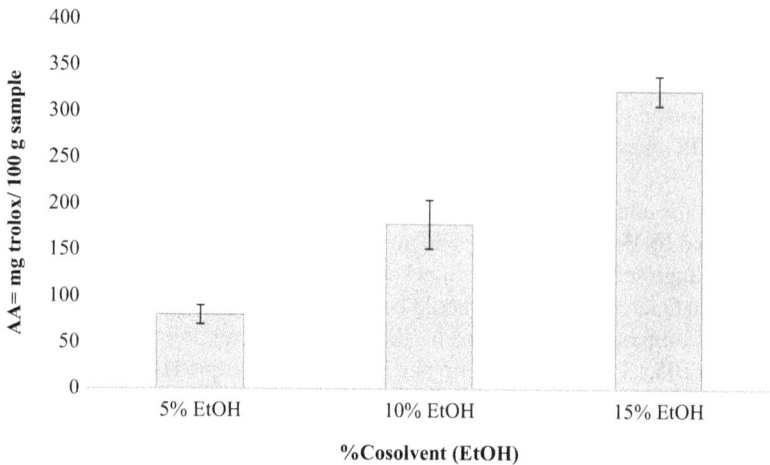

**FIGURE 4.3** Antioxidant activity (mg Trolox/100 g sample) based on EtOH concentrations of 5%, 10%, and 15% (v/v).

*Note:* Error bars represent standard deviation bars.

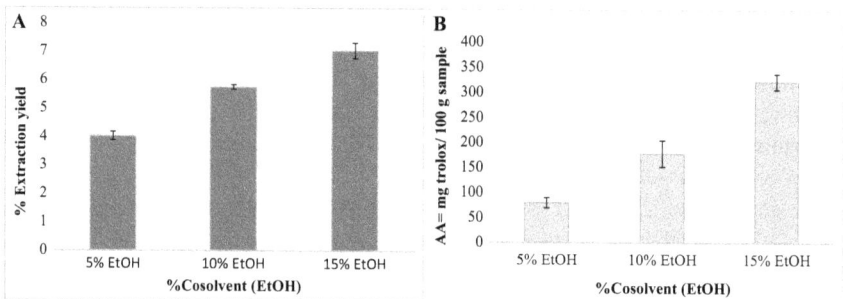

**FIGURE 4.4** Ratio of yield (A) vs. antioxidant activity (AA) (B) based on EtOH concentrations of 5%, 10%, and 15% (v/v).

*Note:* Error bars represent standard deviation bars.

100 g of dry sample (Da Porto et al., 2014). This demonstrates better results and an optimal operative percentage by adding a modifier vs. other methods.

## YIELD AND ANTIOXIDANT ACTIVITY

The yield of Isabella grape pomace extraction (%) and antioxidant activity (mg Trolox/100 g sample) achieved by supercritical fluid extraction with $CO_2$ added with 5%, 10%, and 15% EtOH as co-solvent (60°C, 31.7 MPa, $CO_2$ 5000 mL/min; 0.3, 0.6, and 0.9 mL/min EtOH, 5 hours) is shown in Figure 4.4.

The best extraction yield is directly proportional to the highest antioxidant activity obtained using 15% ethanol. A correlation exists as the lowest conditions with the lowest results are applied as antioxidant capacity and yield % increase as the modifier percentage does due to the use of a polar co-solvent (EtOH) that favors the results, as shown in the graph. The same behavior is reported for other varieties of fruits such as black mashua extracts, where antioxidant capacity is related to the extraction yield with a significant correlation coefficient of 0.712 estimated (Aguirre, 2019).

## OIL OXIDATIVE STABILITY (OOS) THROUGH RANCIMAT

The results for oxidative stability of soybean oil OOS enriched with the extracts obtained are shown in Figure 4.5. Most extracts generated an increased induction time (IT) compared to the control (4.6 h). However, no significant induction time regarding the control was demonstrated by using the ANOVA statistical method. TBHQ presented the highest IT, although it had the same conditions as the extract. Despite this, the capacity of the natural additive vs. the synthetic additive was sought to be showed at the laboratory level in order to learn the extent of protection that could be achieved.

The induction periods given by the Rancimat apparatus show that for the CT (control—extract-free soybean oil) the length is shorter (4.6 hours) compared to oil samples spiked with extract at 5%, 10%, and 15%. A result of increased time is observed due to the increase of modifier although according to the results obtained by the ANOVA statistical method that establish $\alpha = 0.05$ as a significance level no favorable IT for the extracts is proven since $p$ values of 0.70, F of 0.49, and 11 degrees of freedom were reported, discriminating the hypothesis claiming that the results obtained had a protective factor. These results could be due to the effects of the concentration at which the extract is found in oil as a significant variable. The experiments performed could be modified in terms of such a variable as only one concentration was used for every test during the study. Therefore, the likelihood of conducting the experiments with different amounts of extract remains open because

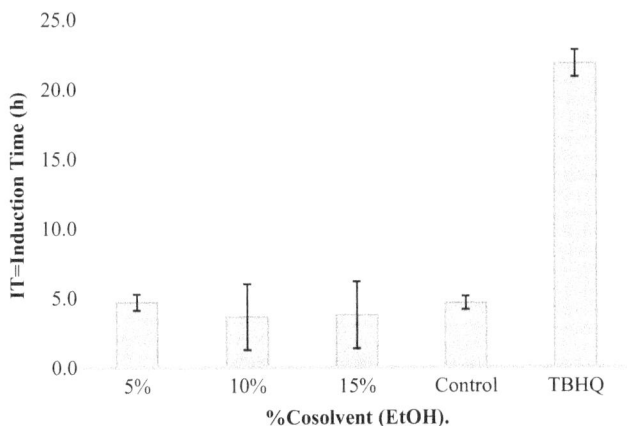

**FIGURE 4.5** Induction time (IT) of extract-spiked soybean oil.

*Note:* Error bars represent standard deviation bars.

a high phenol and flavonoid content was evidenced. Consequently, future laboratory-scale investigations are intended to demonstrate a protective factor as additives for extracts containing a high phenol and flavonoid content. Studies have demonstrated that the Rancimat method using grape extracts indeed has a favorable antioxidant protective quality for the oil, as shown by Lafka et al. in 2007 based on extracts obtained from red wine cellar residues using the Rancimat method on sunflower oil. The analyses showed higher antioxidant activity in ethanol extracts (40–240 ppm) with an induction time of 15.27 hours and a protective factor of 2.05, unlike synthetic antioxidants (BHT, ascorbyl palmitate) and vitamin E that showed induction times of between 9.20 h to 10.23 h, with a protective factor of 1.23–1.37 (Lafka et al., 2007). Collectively, these results mean that extracts from wine-making residues provide optimal protection against oil oxidation, which in turn is conditioned by the amount of extract added to the apparatus.

Continuing with the studies carried out on wine-making residues using the Rancimat apparatus, a significant antioxidant capacity regardless of the type of oil used has been demonstrated for extracts in the laboratory. Nevertheless, studies where the issue has been further addressed for using grape extract oil as an ingredient for the food industry are available as shown in the study conducted by Shirahigue and collaborators in 2010 on chicken meat cooked using Isabella grape extract (IGE) and Niagara grape extract (NGE) as antioxidants. The increased extract (400 µg GAE/mL) improving the soybean oil oxidative stability (30 g) was shown, and Isabella and Niagara grape extract antioxidants additionally displayed similarities with results obtained for TBH (400 µg/mL), with an induction time of 2.35 h for the control, and 4.21 h and 3.73 h for IGE and NGE, respectively. On the other hand, TBH had an IT of 3.94 h (Shirahigue et al., 2010) as opposed to the analyses carried out by Samah et al. in 2012 for other varieties of grape (red, white, and black) where the oxidative stability index (OSI) of sunflower oil added with the synthetic antioxidant TBH showed a better result (11.9) than extracts from grape husk at 100°C. Despite this, results obtained with these extracts were good if the concentration was low; at 200 ppm the OSI ranged between 12.2–12.9, although at 400 ppm it was 12.2–11.4 (Samah et al., 2012). The data show that the antioxidant activity of extracts is due to the properties in grape residues, the method used to obtain the extracts, and the concentration applied to each fat or oil.

## CONCLUSIONS

SFE-$CO_2$ with EtOH as co-solvent gives better results compared to other extracts obtained using SFE-$CO_2$ with no modifier. The yield for these extracts is higher, with a greater content of phenolic compounds and high antioxidant activity. These properties may be improved by increasing certain extraction variables such as the co-solvent percentage, pressure, and temperature. Extraction conditions for this study were a 15% EtOH modifier with a flow rate of 0.9 mL/min, a $CO_2$ flow rate of 5 L/min, a temperature of 60°C, and a pressure of 31.7 MPa, where the best extraction yield and increased TPC, TFC, and antioxidant activity were obtained. Grape pomace extracts applied as antioxidants to soybean oil showed a protective effect against oxidation because of the higher induction times displayed as compared to the control.

Results point at the extracts obtained as potential suggested alternatives to replace synthetic additives as they are a source of health-promoting bioactive compounds. Exploitation of by-products from the wine-making industry under the green extraction technique SFE-CO$_2$ with EtOH as a co-solvent represents another method to use waste-generating products of high value and quality added, with no disturbance of soil and water resources due to the inadequate disposal of waste or if extraction techniques that contaminate the environment and affect human health are used.

## GLOSSARY

**Antioxidant activity:** Ability of a substance to inhibit oxidative degradation (Londoño, 2012).

**By-product:** Product obtained from a productive process susceptible to economic valorization, but less important than the main product (Wolters Kluwer, n.d.).

**Co-solvent:** Chemical substance used in small quantities in order to improve the effectiveness of a primary solvent in a chemical process (Schlumberger, n.d.).

**Flavonoids:** These bioactive compounds are polyphenol secondary metabolites that usually present a ketone group and yellow color pigments in their structure (Ramírez, 2020).

**Grape pomace:** Solid residue obtained after extraction of grape juice that represents the wine-making industry's major by-product (Díaz, 2009).

**Green technology:** Design of solutions based on eco-efficiency as they guarantee manufacturing and operating safety while reducing any environmental impact (Altamira, 2017).

**Modifier:** Component of the organic phase with specific functions at optimizing the extraction process rendering it more efficiency (Durand, 2014).

**Optimal variables:** Favorable conditions that allow obtaining the best results.

**Oxidative induction time:** Time elapsed until high-speed oxidation takes place (Metrohm, n.d.).

**Oxidative stability:** Chemical reaction generated when oil and oxygen are combined (Noria, 2014).

**Phenolic compounds:** Popular name for hydroxybenzene. Phenolic compounds act as phytoalexins (wounded plants secrete phenols to defend against potential fungal or bacterial attacks) and contribute to pigmentation of many parts of the plant (Gimeno, 2004).

**Supercritical fluid:** A state wherein matter is compressible and behaves as gas. However, the typical density of a supercritical fluid is that of a liquid and, therefore, it has the characteristic dissolution potential of a liquid (De Castro et al., 2012).

## ACKNOWLEDGMENTS

To Universidad Libre for financial support, to the Universidad Nacional de Colombia Supercritical Fluids Laboratory for providing their facilities, to the companies Casa Grajales and Duquesa S.A. for providing the raw material (grape pomace and soybean oil) required to conduct the study.

## REFERENCES

Aguirre Huayhua, L. L. (2019). Evaluación de presión, temperatura y cosolvente en el rendimiento y actividad antioxidante de antocianinas extraídas de mashua negra por fluidos supercríticos (Tesis de posgrado). Universidad Nacional del Centro del Perú. Huancayo, Perú.

Ahn, J; Grun, I; & Mustapha, A. (2007). Effects of plant extracts on microbial growth, color change, and lipid oxidation in cooked beef [en linea]. Disponible en internet: https://doi.org/10.1016/j.fm.2006.04.006

Almeida-Doria, R. F., & Regitano-D'Arce, M. A. (2000). Antioxidant activity of rosemary and oregano ethanol extracts in soybean oil under thermal oxidation. Food Science and Technology, 20(2), 197–203.

Altamira, L. M. D. R. S. (2017). Sustentabilidad y Tecnología Verde. Recuperado de www.gestiopolis.com/wp-content/uploads/2017/03/sustentabilidad-ytecnologia-verde-mexico.pdf

Anbudhasan, P., Surendraraj, A., Karkuzhali, S., & Sathishkumaran, P. (2014). Natural antioxidants and its benefits. International Journal of Food and Nutritional Sciences, 3(6), 225.

Barragan, Blanca & Azucena, Tellez-Díaz & Adriana, Lagua-Trinidad. (2008). Utilización De Residuos Agroindustriales. Revista Sistemas Ambientales. 2. 44–50.

Barros, L., De Mio, L., Biasi, L., Di Profio, F., & Reynolds, A. (2014). Use of HPLC for characterization of sugar and phenolic compounds in Vitis labrusca juice. IDESIA, 32(2), 89–94. Recuperado en: https://scielo.conicyt.cl/pdf/idesia/v32n2/art12.pdf

Buelvas, L. (2018). Valorización y modelamiento de la extracción de aceites a partir de la semilla de mango (Mangifera indica L.) utilizando técnicas no convencionales. (Magister en Ingeniería Química). Universidad Nacional de Colombia, Bogotá D.C.

Calvo, A., Morante, J., Plánder, S., & Székely, E. (2017). Fractionation of biologically active components of grape seed (Vitis vinifera) by supercritical fluid extraction. Acta Alimentaria, 46(1), 27–34

Casas, E., Faraldi, M., Bildstein, M.(2008) Evaluación y difusión de las estrategias para la extracción de compuestos bioactivos de residuos del procesado del tomate, de la aceituna y de la uva. Recuperado de www.ainia.es/html/portal_del_asociado/uva.pdf

Castejón, N., Luna, P., & Señoráns, F. J. (2018). Alternative oil extraction methods from Echium plantagineum L. seeds using advanced techniques and green solvents. Food Chemistry, 244, 75–82. https://doi.org/10.1016/J.FOODCHEM.2017.10.014

Chang, C. C., Yang, M. H., Wen, H. M., & Chern, J. C. (2002). Estimation of total flavonoid content in propolis by two complementary colorimetric methods. Journal of food and drug analysis, 10(3).

Cuellar, R. (2017). Diseño de la Automatización para una Planta Piloto de Extracción por Fluido Supercrítico Utilizando $CO_2$ como Solvente. (Tesis de posgrado). Pontificia Universidad Católica Del Perú, Perú.

Da Porto, C., Decorti, D., & Natolino, A. (2014). Water and ethanol as co-solvent in supercritical fluid extraction of proanthocyanidins from grape pomace: A comparison and a proposal. The Journal of Supercritical Fluids, 87, 1–8.

De Campos, L. M., Leimann, F. V., Pedrosa, R. C., & Ferreira, S. R. (2008). Free radical scavenging of grape pomace extracts from Cabernet sauvignon (Vitis vinifera). Bioresource Technology, 99(17), 8413–8420.

De Castro, M. D. L., Valcárcel, M., && Tena, M. T. (2012). Analytical supercritical fluid extraction. Springer Science & Business Media.

Díaz Sánchez, A. B. (2009). Reciclado del orujo de uva como medio sólido de fermentación para la producción de enzimas hidrolíticas de interés industrial. Universidad de Cádiz.

Ding, M., & Zou, J. (2012). Rapid micropreparation procedure for the gas chromatographic-mass spectrometric determination of BHT, BHA and TBHQ in edible oils. Food Chemistry, 131(3), 1051–1055.

Durand Venegas, J. A. (2014). Optimización en la etapa de extracción por solventes en base al tipo de diluyente en el orgánico, en una unidad minera. Escuela Académica Profesional de Ingeniería Metalúrgica, Tacna, Perú.

Espinosa, F. (2017). Comparación de la Actividad Antioxidante del Extracto Etanolico del Esparrago (Asparragus officinalis), Tocoferol y TBHQ utilizando el método Rancimat (Tesis de pregrado). Universidad de Sonora, Sonora.

Farías-Campomanes, A. M., Rostagno, M. A., & Meireles, M. A. A. (2013). Production of polyphenol extracts from grape bagasse using supercritical fluids: Yield, extract composition and economic evaluation. The Journal of Supercritical Fluids, 77, 70–78.

Gavilan Guillen, N. J. (2016). Efecto de presión y temperatura en extracción por CO2-supercrítico y etanol en capsaicinoides de venas de ají panca (Capsicum chinense), (tesis de pregrado). Universidad Nacional del Centro del Perú, Huancayo, Perú.

Ghafoor, K., AL-Juhaimi, F. Y., & Choi, Y. H. (2012). Supercritical fluid extraction of phenolic compounds and antioxidants from grape (Vitis labrusca B.) seeds. Plant Foods for Human Nutrition, 67(4), 407–414

Ghafoor, K., Park, J., & Choi, Y. H. (2010). Optimization of supercritical fluid extraction of bioactive compounds from grape (Vitis labrusca B.) peel by using response surface methodology. Innovative Food Science & Emerging Technologies, 11(3), 485–490.

Gimeno, E. (2004). Compuestos fenólicos. Un análisis de sus beneficios para la salud. Offarm, vol. 23. Núm. 6. páginas 80–84.

Guerrero, R. & Valenzuela, L. (2011). Agroindustria y medio ambiente. Trilogía, Ciencia Tecnología Sociedad, 23(33):63–83.

Herrera, S. Y., Zampini, I. C., D'Almeida, R., Boguetti, H., & Isla, M. I. (2007). Extracción con fluido supercrítico de compuestos con capacidad Antioxidante de Baccharis incarum: comparación con métodos convencionales. Boletín Latinoamericano y del Caribe de Plantas Medicinales y Aromáticas, 6(5), 250–251.

Khezerlou, A., Akhlaghi, A. pouya, Alizadeh, A. M., Dehghan, P., & Maleki, P. (2022). Alarming impact of the excessive use of tert-butylhydroquinone in food products: A narrative review. Toxicology Reports, 9, 1066–1075. https://doi.org/10.1016/J.TOXREP.2022.04.027

Lafka, T. I., Sinanoglou, V., & Lazos, E. S. (2007). On the extraction and antioxidant activity of phenolic compounds from winery wastes. Food Chemistry, 104(3), 1206–1214.

Londoño Londoño,J. (2012). Antioxidantes: importancia biológica y métodos para medir su actividad. En Desarrollo y transversalidad serie Lasallista Investigación y Ciencia. Corporación Unversitaria Lasallista.

Luque de Castro, M., Valcárcel Cases, M. and Tena, M. (1993). Extracción Con Fluidos Supercríticos En El Proceso Analítico. Barcelona: Reverté, p. 60.

Magalhães, L. M., Barreiros, L., Maia, M. A., Reis, S., & Segundo, M. A. (2012). Rapid assessment of endpoint antioxidant capacity of red wines through microchemical methods using a kinetic matching approach. Talanta, 97, 473–483.

Martínez, C, & Ceballos, C. (2012). Determinación de actividad antioxidante en aceite de semillas de uva isabella (Vitis labrusca) extraido con $CO_2$ supercrítico. (tesis de pregrado). Universidad del Valle, Santiago de Cali, Colombia.

Metrohm. (s.f.). Rancimat. Recuperado de www.metrohm.com/es/productos/medicion-de-la-estabilidad/rancimat/#

Michalak, M. (2022). Plant-Derived Antioxidants: Significance in Skin Health and the Ageing Process. International Journal of Molecular Sciences, 23(2). https://doi.org/10.3390/ijms23020585

Noria. (2014). La importancia de la estabilidad a la oxidación. Recuperado de https://noria. mx/lublearn/la-importancia-de-la-estabilidad-a-la-oxidacion/#:~:text=La%20esta bilidad%20a%20la%20oxidaci%C3%B3n%20es%20una%20reacci%C3%B3n%20 qu%C3%ADmica%20que,oxidaci%C3%B3n%20incrementa%20con%20el%20tiempo.

Otero-Pareja, M. J., Casas, L., Fernández-Ponce, M. T., Mantell, C., & Ossa, E. J. (2015). Green extraction of antioxidants from different varieties of red grape pomace. Molecules, 20(6), 9686–9702.

Palacios, N. (2013). Un método para generar mezclas CO2 + etanol en estado supercrítico. Investigaciones en Facultades de Ingeniería del NOA ISSN N° 1853-7871. Recuperado de https://fcf.unse.edu.ar/archivos/publicaciones/codinoa-2013/trabajos/tecnologicas/75-palacios.pdf

Pateiro, M., Barba, F. J., Domínguez, R., Sant'Ana, A. S., Mousavi Khaneghah, A., Gavahian, M., Lorenzo, J. M. (2018). Essential oils as natural additives to prevent oxidation reactions in meat and meat products: A review. Food Research International, 113, 156–166.

Pinelo, M., Arnous, A., & Meyer, A. (2006). Upgrading of grape skins: Significance of plant cell-wall structural components and extraction techniques for phenol release. Trends in Food Science & Technology, 17(11), 579–590. doi: 10.1016/j.tifs.2006.05.003.

Rahman, M. M., Rahaman, M. S., Islam, M. R., Rahman, F., Mithi, F. M., Alqahtani, T., Almikhlafi, M. A., Alghamdi, S. Q., Alruwaili, A. S., Hossain, M. S., Ahmed, M., Das, R., Emran, T. Bin, & Uddin, M. S. (2022). Role of phenolic compounds in human disease: Current knowledge and future prospects. In Molecules (Vol. 27, Issue 1). MDPI. https://doi.org/10.3390/molecules27010233

Ramírez, S. (2019). Cuantificacion por HPLC de Trans-Resveratrol en la Uva Isabella (Vitis Labrusca) Cultivadas en la Unión Valle del Cauca (tesis de pregrado). Universidad Santiago de Cali, Cali, Colombia.

Ramírez, C. (2020). Evaluación de la Extracción de Flavonoides a Partir de la Cáscara de Naranja. Proyecto Integral de Grado para optar al título de Ingeniera Química. Fundación Universidad de América, Bogotá, Colombia.

Salcedo, A. V. R., González, A. F. R., & Alzate, C. A. C. (2017). Obtención de compuestos fenólicos a partir de residuos de uva Isabella (vitis labrusca). Ingresar a La Revista, 15(2), 79–72.

Samah, M., Soltan, S. S., Khaled, A., & Hoda, M. H. (2012). Phenolic compounds and anti-oxidant activity of white, red, black grape skin and white grape seeds. Life Sci J, 9(4), 3464–74.

Santos, K. A., Frohlich, P. C., Hoscheid, J., Tiuman, T. S., Gonçalves, J. E., Cardozo-Filho, L., & da Silva, E. A. (2017). Candeia (Eremanthus erythroppapus) oil extraction using supercritical CO$_2$ with ethanol and ethyl acetate cosolvents. The Journal of Supercritical Fluids, 128, 323–330

Sapkale, G. N., Patil, S. M., Surwase, U. S., & Bhatbhage, P. K. (2010). Supercritical fluid extraction. Int. J. Chem. Sci, 8(2), 729–743.

Segura, C., Guerrero, C., Posada, E., Mojica, J., & Mora, W. P. (2015). Caracterización de residuos de la industria vinícola del valle de Sáchica con potencial nutricional para su aprovechamiento después del proceso agroindustrial. Investigación Bogotá, 1–6.

Shirahigue, L. D., Plata-Oviedo, M., De Alencar, S. M., D'Arce, M. A. B. R., De Souza Vieira, T. M. F., Oldoni, T. L. C., & Contreras-Castillo, C. J. (2010). Wine industry residue as antioxidant in cooked chicken meat. International Journal of Food Science & Technology, 45(5), 863–870.

Schlimberger. (s.f.). Oilfield Glossary en Español. Recuperado de https://glossary.oilfield.slb. com/es/terms/c/cosolvent

Soler Fernández, L. (2017). Valorización de los residuos procedentes de una producciónvitivinícola (Doctoraldissertation). https://riunet.upv.es/bitstream/handle/10251/91267/TFG_SolerFernandezLucia.pdf?sequence=3

Tournour, H. H., Segundo, M. A., Magalhães, L. M., Barreiros, L., Queiroz, J., & Cunha, L. M. (2015). Valorization of grape pomace: Extraction of bioactive phenolics with antioxidant properties. Industrial Crops and Products, 74,397–406.

Vatai, T., Škerget, M., & Knez, Ž. (2009). Extraction of phenolic compounds from elder berry and different grape pomace varieties using organic solvents and/or supercritical carbon dioxide. Journal of Food Engineering, 90(2), 246–254.

Velásquez, A. M. (24 noviembre del 2008). La tecnología de fluidos supercríticos, un proceso limpio para el sector industrial. Revista Producción más limpia. Recuperado de http://lasallista.edu.co/fxcul/media/pdf//revistalimpia/vol3n2/88-97.pdf

Waterhouse, A. L. (2002). Determination of total phenolics. Current Protocols in Food Analytical Chemistry, 6(1), I1–1.

Wolters Kluwer. (s.f.). Subproductos. Recuperado de https://guiasjuridicas.wolterskluwer.es/Content/Documento.aspx?params=H4sIAAAAAAAEAMtMSbF1jTAAASNTM2NTtbLUouLM_DxbIwMDS0NDQ3OQQGZapUt-ckhlQaptWmJOcSoAYfS9BDUAAAA=WKE

Yalcin, H., Kavuncuoglu, H., Ekici, L., & Sagdic, O. (2017). Determination of fatty acid composition, volatile components, physico-chemical and bioactive properties of grape (*Vitis vinifera*) seed and seed oil. Journal of Food Processing and Preservation, 41(2), e12854.

Zulkafli, Z. D., Wang, H., Miyashita, F., Utsumi, N., & Tamura, K. (2014). Cosolvent-modified supercritical carbon dioxide extraction of phenolic compounds from bamboo leaves (*Sasa palmata*). The Journal of Supercritical Fluids, 94, 123–129. doi: 10.1016/j.supflu.2014.07.008

# 5 Obtaining Potentially Biologically Active Extracts from Isabella Grape Pomace Using Pressurized Liquids and Evaluation of Functional Properties

*María Alejandra Castañeda Muñoz,*
*Liced Alejandra Basto Gómez,*
*Patricia Joyce Pamela Zorro Mateus, and*
*Henry Isaac Castro Vargas*

Large quantities of biomass subjected to an industrial process in order to obtain a product are produced in Colombia. During this process some raw materials are not fully exploited, they can become high value-added waste which they undergo transformation for greater exploitation (García Morales et al., 2014). Some by-products of fruits, vegetables, or algae, among others, are a source of bioactive compounds with antioxidant capacity (Chandrasekaran, 2012). The agri-business of Isabella grape (*Vitis labrusca* L.) for wine production is associated with the generation of large quantities of by-products whose level of utilization is reduced over time due to rapid decomposition. Upon disposal, such by-products represent a significant problem for the environment. However, these by-products may be exploited in different ways from food for animals or humans, or obtaining compost, to more complex methods including, but not limited to, the extraction of oils and additives, and the production of bioethanol and dyes (Yepes et al., 2008). An example of this is grape pomace, which is an agri-business by-product from this fruit regarded as a source of antioxidant compounds (Segura et al., 2015; Lucarini et al., 2018) bearing a large amount of phenolic compounds (Makris et al., 2006; Prieto et al., 2011; Antoniolli et al., 2015; Vargas and Vargas et al., 2019) such as anthocyanins, classified among the flavonoids

DOI: 10.1201/9781003391593-5

**65**

(Hid Cadena et al., 2010) which give rise to the featured dark purple color of this fruit (Creasy & Creasy, 2009).

Considering the claims above, adding the negative environmental effects triggered by the wine industry grape waste, and taking into account that such waste causes a progressive degradation of the environment due to the high amount, composition, and accumulation of these remains, the reuse and valorization of grape residues is necessary (i.e., undergoing extraction processes) in order to take advantage of the components and properties existing therein. One extraction technique is extraction using pressurized liquids (PLE) which may be optimal for recovery of different compounds in used plant material and brings about several advantages for the industry as well. Pressurized liquid extraction is the most inexpensive and ecology-friendly alternative having reduced extraction times, collecting higher quality extracts, and good procedural reproducibility (Becerra et al., 2017). Furthermore, PLE meets the market environmental requirements as environmentally friendly solvents are implemented. The interest in using this technique has been accordingly increasing in recent years.

For PLE extraction, solvents at high temperatures and pressures keeping the solvent in a liquid state throughout the extraction process are employed. In PLE, different variables such as temperature, solvent type, extraction flow, and pressure are interrelated (Turner & Waldebäck, 2013; Alvarez-Rivera et al., 2019). These parameters, if modified, alter the characteristics of the extract obtained; therefore, the aim of this research was to optimize the variables associated with PLE extraction based on the recovery of phenolic compounds and their antioxidant properties to hence chemically characterize the extracts obtained regarding the total content of phenolic compounds and to test the antioxidant properties for each extract.

Considering the above and to give added value to residues generated in the production of Isabella grape (*Vitis labrusca* L.) wine a process for recovery of phenolic antioxidants from by-products created in the wine-making industry was carried out using PLE.

## THEORETICAL FRAMEWORK

Agro-industrial residues in close contact with the soil in particular cause desertification, pore saturation, acidity, and soil surface erosion (García Morales et al., 2014). At the same time, the accumulation and decomposition of these residues generate air pollution due to the release of bad odors and greenhouse gases ($CO_2$, $CH_4$, $NO_2$). Likewise, agro-industrial residues impact water bodies given the large amount of waste, many of which contain pesticides that are carried by runoff and currents, thus causing potential acidification and contamination of resources (Mejías et al., 2016).

The grape in all its varieties is an agro-industrial product with over 75 million tons produced worldwide, with approximately 50% of the total harvest used in wine-making (FAO and OIV, 2016), therefore grape residue is a significant source of agro-industrial waste, including: pomace (remnants of pressed grapes), lees (precipitated during fermentation and maturation), and vinasse (washing waters). According to the International Organisation of Vine and Wine, in Colombia, 32,298 tons of grapes were produced in 2021 (International Organisation of Vine and Wine [OIV], 2022) with the

Valle del Cauca region contributing 64% of the national production (Agronet, 2018). Therefore, the increased supply and demand for wine production grapes in Colombia has significantly increased the amount of agro-industrial waste and also has led to the environmental disturbances mentioned above. However, significant anti-oxidant and anti-microbial properties, particularly in the gallic acid obtained from grape peels, have been reported, with features such as the total content of flavonoids, catechins, resveratrol, and stilbenes, evidence of a high phenolic content such as the already mentioned anthocyanin standing out (Figures 5.1 and 5.2) (Katalinić et al., 2010; Dávila et al., 2017).

In fact, the benefits mentioned above confirm the importance of reusing agro-industrial waste, in this case Isabella grape waste, paving the way for processes involving new uses of the by-products collected from industrial processes. In order to pursue this line of environmental preservation and care, the likelihood is increased of taking advantage of these by-products by means of green technologies such as the PLE method used in this research, which follows the "six principles of green

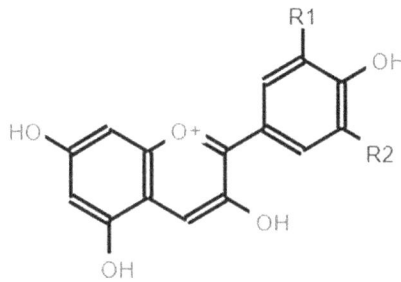

**FIGURE 5.1** General structure of anthocyanin.

Adapted from Cambios en contenido de compuestos fenólicos y color de extractos de Jamaica (Hibiscus sabdariffa) sometidos a calentamiento con energía de microondas, by R. Hid et al., 2010, Universidad de Alicante.

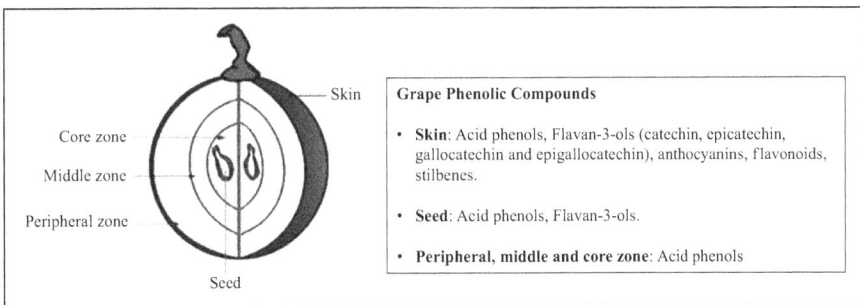

**FIGURE 5.2** Structure and phenolic compounds of grape.

Adapted from Handbook of Grape Processing by Products: Sustainable Solutions (p. 33), by Dávila et al., 2017, Academic Press.

extraction for natural products" proposed by Chemat in 2012 involving: innovation by selection of renewable resources, use of alternative solvents in lieu of conventional solvents mainly water or agro-solvents, reduction of energy consumption using innovative technologies, production of by-products instead of waste, reduction of unitary operations promoting process automation, and collection of contaminant-free, non-denatured biodegradable extracts.

First, in summary, extraction is the operation of transferring matter based on the dissolution of the sample components in a selective solvent (Costa López, 2004). PLE is a method where solvents are below their critical point at high temperatures and pressures, causing the solvents to remain in a liquid state even above their boiling point, making it possible to obtain extracts from solid samples (Rivas, 2007; Picó, 2017; Kultys & Kurek, 2022). In this way, the target compounds (i.e., from grape pomace) are quickly and efficiently extracted, recovering a larger amount of elements. The PLE process is efficient and recommended for the extraction of bioactive compounds due to the solubility and desorption kinetics of the target analytes (Tamires Vitor et al., 2020) and has been used for removal of soil, sediment, and animal tissue contaminants. Nonetheless, PLE applications have now progressed to the field of food and pharmaceutical products (Becerra et al., 2017), being mainly used to extract antioxidants such as polyphenols (Alvarez-Rivera et al., 2019).

Currently, PLE is an alternative, low-cost, green (a decreased amount of waste generated) extraction technique compared to conventional forms of extraction such as maceration, Soxhlet, liquid–liquid extraction, solid–liquid extraction, or ultrasound (Han et al., 2011; Wianowska & Gil, 2019). Thus, according to Panja (2018), the advantages of PLE can be encompassed in a larger amount of phenolic compounds recovered, energy savings in the process, and low cost of the solvents used, in addition to being environmentally friendly and non-toxic. Likewise, the time required in the process is reduced compared to other techniques (Do et al., 2013) with PLE being more efficient than other types of extraction (Santos et al., 2013).

Phase liquid extraction consists of a pump-coupled solvent reservoir that transfers the solvent to an extraction cell containing the sample on a thimble. The system is fitted with a valve to maintain the pressure inside. The pump transfers the solvent to the cell continuously until filled to the top. Pressure and subsequently temperature are increased in the system, thus allowing the solvent to remain in a liquid state. Finally, the solvent flows through the sample into the cell so the extract is further collected. Once the process is completed the cell is purged using the same solvent mixture to remove any residues. Collected extract is then ready to be concentrated.

On the other hand, there are two main PLE configurations: static and dynamic. In the dynamic configuration, solvent is pumped continuously through the sample with constant flow, while in the static mode, the sample is enclosed in a stainless steel container filled with an extraction solvent for a certain amount of time (Mustafa & Turner, 2011). It is worth noting that extraction efficiency may increase in the dynamic mode (Picó, 2017) because, under static conditions, achievement of the analyte concentration equilibrium between the sample and solvent is likely leading to an idle process and not allowing higher efficiency. In the dynamic mode, on the other hand, the equilibrium is shifted by continuously introducing solvent to the sample, allowing a better yield (Herrero et al., 2013).

One of the most important components within the PLE process is optimization of the extraction parameters such as temperature, pressure, type of solvent, percentage of each solvent if it is a mixture, among other aspects. First, temperature is considered to beone of the most important parameters in PLE because high temperatures decrease the dielectric constant and modify solvent polarity so the latter can approximate or match the polarity of the bioactive compounds to be extracted. Likewise, the increase in temperature decreases the viscosity of a liquid solvent, thus improving penetration into the sample (Dunford et al., 2010; Miron et al., 2011). For example, in water, the dielectric constant decreases at high temperatures, resulting in a solvent with ethanol- and methanol-like properties at room temperature (Yang et al., 1995; Carr et al., 2011). Similarly, by using liquids at high temperature and pressure the liquid diffusion coefficient is increased and surface tension is decreased, accelerating the process and requiring less solvent for extraction (Turner & Waldebäck, 2013). Simply put, the high temperature used for PLE makes solvents and analytes more related, generating a higher yield extraction process.

Temperatures used in PLE range from 50°C to 200°C (Pasrija & Anandharamakrishnan, 2015). However, for extraction of phenolic compounds it is important not to use such elevated temperatures as these may contribute to degradation of some compounds. For thermolabile phenolic compounds the temperature used usually ranges from 40°C to 60°C, whereas for thermostable phenolic compounds temperatures of from 75°C to 220°C are used (Alvarez-Rivera et al., 2019) although it is recommended to use temperatures lower than 160°C to avoid reduction of flavonoids and ascorbic acid, among other compounds (Howard & Pandjaitan, 2008).

In PLE, the interaction between high temperatures and type of solvent facilitates compound solubility, rate of solvent diffusion, and mass transfer (Poole, 2020). Consequently, fewer solvents are used and less time is required to perform an extraction. The PLE extraction technique allows and makes flexible the use of different types of solvents selected according to the nature of the compound to be extracted. For lipophilic and non-polar compounds, non-polar and/or semi-polar solvents such as hexane and pentane are used. On the contrary, polar solvents such as acetonitrile, methanol, or water are used to extract polar compounds (Becerra et al., 2017). It should be mentioned that in PLE, water in the first place and other renewable solvents such as ethanol or isopropanol are considered "green" solvents (Chemat & Strube, 2015). It is worth noting that the extraction efficiency may be improved by using a mixture of solvents, with hydro-ethanol mixtures being the most suitable for phenolic compounds (Carabias-Martínez et al., 2005; Picó, 2017; Alvarez-Rivera et al., 2019).

As for pressure in PLE, it is limited to play a role in keeping the solvent in a liquid state even at temperatures above its boiling point. Pressures between 7 MPa and 14 MPa (1000–2000 psi) are usually applied without affecting the efficiency or characteristics of the extract obtained (Richter & Raynie, 2012; Plaza & Turner, 2015). Finally, the extraction time refers to the length of time a sample is in contact with the solvent where the effect of this parameter depends on the PLE extraction mode used, with the static configuration time being up to 20 minutes (Herrero et al., 2013) and dynamic configuration where times may be longer. For any type of PLE extraction (static or dynamic) it is important to ensure that the analytes have sufficient time to achieve the highest mass transfer (Richter & Raynie, 2012).

Sample particle size is very important, and treatment should be implemented in order to reduce it as mass transfer is influenced by the particle size and having a larger contact surface is paramount. In addition, by decreasing the particle size it is possible to break down cell walls while improving diffusion of the analyte (Richter & Raynie, 2012). Similarly, removing moisture from the sample is important as it increases the extraction efficiency (Carabias-Martínez et al., 2005).

## METHODOLOGY

### SAMPLE PREPARATION

Agro-industrial residues of Isabella grape (*Vitis labrusca* L.) supplied by the liquor maker Casa Grajales located in the municipality of La Unión, Valle del Cauca, were used. This plant material was subjected to a drying process and subsequent manual cleaning to remove impurities and other residues, branches or rachis were separated, and peels and seeds were kept as working material (Figure 5.3). The sample was immediately milled using a traditional mill (Corona brand) and finally sieved through a screen (Standard ASTME-11 brand) for selection of a 0.35 mm particle size.

### PLE EXTRACTION

Extracts from Isabella grape by-products were obtained through pressurized liquid extraction using a dynamic mode configuration by means of a solvent (ethanol:water) supplying pump to the extraction cell fitted with a thimble that contains the grape pomace sample. The apparatus is equipped with a pressure-regulator valve (BPR) and a heat exchanger to control pressure and temperature, following the procedure described previously (Mustafa & Turner, 2011; Christen & Kaufmann, 2014; Ballesteros, 2015) (Figure 5.4). It is worth mentioning that the PLE equipment used was supplied by the Department of Chemistry's Food Chemistry Research Group (Laboratory 131) of the Universidad Nacional de Colombia.

FIGURE 5.3   Selected and dried pomace.

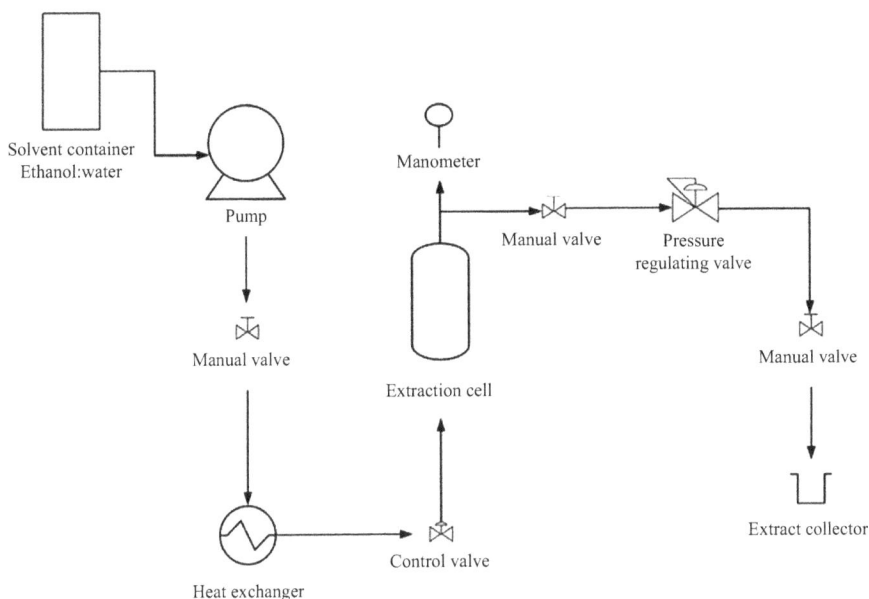

**FIGURE 5.4** Pressurized liquid extraction process.

The tested factors were the extractant phase composition and the extraction temperature, which were determined by a central composite rotable experimental design made up of a 22 factorial design with axial points (α=1.414) indicating the maximum and minimum values of process parameters, in this case temperature and ethanol %, where four repetitions at the central point (50% EtOH and 80°C) are carried out. The points tested for extraction temperature were 51.7°C, 60°C, 80°C, and 108.3°C, and the extractant phase composition of 35.9%, 40%, 50%, 60%, and 64.1% EtOH. As process factors, an extraction time of 2 hours, flow of 5.59 mL/min, pressure of 0.482 MPa, sample amount of 5.02 g (Annex 5.1), and a dynamic extraction configuration mode were set up. Conditions for each extraction in terms of temperature and EtOH % are specified in Table 5.1. In the extraction process, the effect of the process parameters associated with PLE on the variables yield (%) response, total phenol content (TPC), total flavonoid content (TFC), and antioxidant activity analyzed by the DPPH method were determined (see Annex 5.2).

Subsequently, the collected extracts were concentrated to dryness by means of a Buchi R-300 rotary evaporator. Lastly, yield, TPC, TFC, and antioxidant activity were determined.

## TOTAL PHENOL CONTENT

As for the determination of the total phenol content, it was carried out by the Folin-Ciocalteu method proposed by Otto Folin and Vintila Ciocalteu (Singleton et al., 1999). Briefly, a 1 mg/mL gallic acid standard solution stock was prepared and dilutions from the stock using distilled water to obtain different concentrations for the

**TABLE 5.1**
**Conditions for each extraction**

| Assay | D (%EtOH) | T (°C) |
|-------|-----------|--------|
| 1 | 40.0% | 60.0 |
| 2 | 60.0% | 60.0 |
| 3 | 40.0% | 100.0 |
| 4 | 60.0% | 100.0 |
| 5 | 50.0% | 51.7 |
| 6 | 50.0% | 108.3 |
| 7 | 35.9% | 80.0 |
| 8 | 64.1% | 80.0 |
| 9 | 50.0% | 80.0 |

calibration curve were made. On the other hand, a dilution of the extract was prepared adding 20 µL to 1580 µL of distilled water and mixing in 100 µL of pure Folin reagent. The reaction mixture was incubated for 8 minutes in complete darkness and after this incubation time had elapsed 300 µL of 20% $Na_2CO_3$ were added. Mixtures were left to stand in the dark for 2 hours and then transferred to optical cuvettes for reading out each absorbance in a Multiskan Sky 1530 model spectrophotometer (Thermo Fisher Scientific) at a wavelength of 765 nm following the methodology by Magalhães et al. (2012) and Waterhouse (2003) with some modifications. Values are reported as mg GAE/100 g.

## TOTAL FLAVONOID CONTENT

On the other hand, measurement of the total flavonoid content was performed by an aluminum chloride-based colorimetric method. For the calibration curve, 1 mg/mL of quercetin solution was used. The standard solutions were prepared using 100 µL of reconstituted extracts each and mixed with 300 µL of 95% ethanol, 50 µL of aluminum chloride, 50 µL of potassium acetate, and 500 µL of distilled water, then incubated for 45 minutes at room temperature. Absorbance was read by means of a Multiskan Sky 1530 model spectrophotometer (Thermo Fisher Scientific) at 415 nm following the methodology described by Woisky (Woisky & Salatino, 1998) and Chang (Chang et al., 2020). The results are expressed as quercetin equivalents (µg QE/g extract).

## ANTIOXIDANT ACTIVITY

For antioxidant activity testing, the DPPH method was used. Briefly, 150 mL of diluted samples were taken and 150 mL of 50% ethanol and 150µL of Trolox standard solution were added; once 120 minutes of reaction elapsed the absorbance decrease was read at a wavelength of 517 nm (Gaviria Montoya et al., 2012; Magalhães et al., 2012). The results were expressed in values of mg Trolox/100 g of extract.

## STATISTICAL DATA ANALYSIS

All experiments were performed in triplicate. Analysis of the statistical differences for resulting data was carried out using the Statgraphics Centurion XVIII software (2018) where by means of a Pareto diagram and response surface the factor predominantly influencing each of the variables studied is indicated. Each variable yielded a corresponding quadratic equation.

## GEOGRAPHICAL AND LEGAL FRAMEWORK

Biomass used for this research was from Isabella grape pomace supplied by the liquor maker Casa Grajales located in the municipality of La Unión, department of Valle del Cauca, known for the extensive production of high-quality grapes. Regarding the legal framework, no documentation or regulations on recovery, use or exploitation of agro-industrial waste in the national territory are currently available.

## RESULTS AND DISCUSSION

The overall results of extractions and analyses are provided in Table 5.2. The analyses were conducted as mentioned above using Isabella grape (*Vitis labrusca* L.) pomace by means of PLE extraction in dynamic mode where the extraction temperature and extractant phase composition were established using the conditions specified in Table 5.1, values that were taken for temperature and EtOH % as –1.5, –1, 0, 1, and 1.5 in response surface plots, as specified in Table 5.3. Extractions were carried out for 2 hours, a flow of 5.59 mL/min, pressure of 0.482 MPa, and 5.02 g as sample amount (Annex 5.1) to evaluate the extraction yield at this step. Subsequently, analyses of total phenolic compounds content (TPC), total flavonoid content (TFC), and antioxidant activity by DPPH were performed on extracts.

Table 5.2 shows the results of the central composite rotable experiment for Isabella grape extracts obtained by PLE using the conditions shown in Table 5.1, as a result of the each variable optimization using the Statgraphics software where a quadratic model was determined for each variable.

The extraction yield value ascertained for each experiment is found in Table 5.2, where an extraction yield variation between 4.5–25.1%, the extract with the highest yield (25.1%) was that from assay 6 (108.3°C, 50% EtOH), and the lowest value obtained at 51.7°C and 50% EtOH (4.5% yield) are shown. These results may be due to the fact that as temperature is increased the physical-chemical characteristics of the solvent improve, which allows for easier penetration of the solvent, thus increasing performance (Figure 5.5).

In a standardized Pareto diagram (Figure 5.6A) it can be noted that the factor most influencing the extraction yield was temperature, which is in consonance with the results exposed above. For the variable of yield response it was found that the efficiency improves as the temperature increases, as stated by Pico in 2017. As previously elucidated, a high temperature improves the yield since the diffusion rate and solubility of the compounds to be analyzed are increased as shown in the response

**TABLE 5.2**
**Results obtained from the central composite rotable experimental design for Isabella grape extracts using PLE**

| Assay | D (%EtOH) | T (°C) | Yield | Average TPC (mg GAE/ g extract) | Average TPC (mg GAE/ 100 g sample) | Average TFC (µg QE/g extract) | DPPH (mmol average mg Trolox/g extract) |
|---|---|---|---|---|---|---|---|
| 1 | 40.0% | 60.0 | 7.5% | 166.26 | 12.47 | 614.87 | 1.22 |
| 2 | 60.0% | 60.0 | 1.6% | 196.29 | 3.19 | 590.36 | 1.89 |
| 3 | 40.0% | 100.0 | 21.0% | 229.54 | 48.25 | 703.97 | 5.42 |
| 4 | 60.0% | 100.0 | 23.1% | 234.96 | 54.31 | 711.04 | 5.66 |
| 5 | 50.0% | 51.7 | 4.5% | 181.88 | 8.18 | 651.76 | 1.37 |
| 6 | 50.0% | 108.3 | 25.1% | 228.85 | 57.53 | 756.94 | 4.79 |
| 7 | 35.9% | 80.0 | 9.5% | 190.21 | 18.07 | 648.54 | 1.67 |
| 8 | 64.1% | 80.0 | 8.0% | 203.97 | 16.38 | 680.38 | 2.32 |
| 9 | 50.0% | 80.0 | 6.05% | 165.87 | 10.03 | 618.46 | 1.17 |

**TABLE 5.3**
**Values taken for response surface plots**

| | −1.5 | −1 | 0 | +1 | +1.5 |
|---|---|---|---|---|---|
| D (% EtOH) | 35.9 | 40 | 50 | 60 | 64.1 |
| T (°C) | 51.7 | 60 | 80 | 100 | 108.3 |

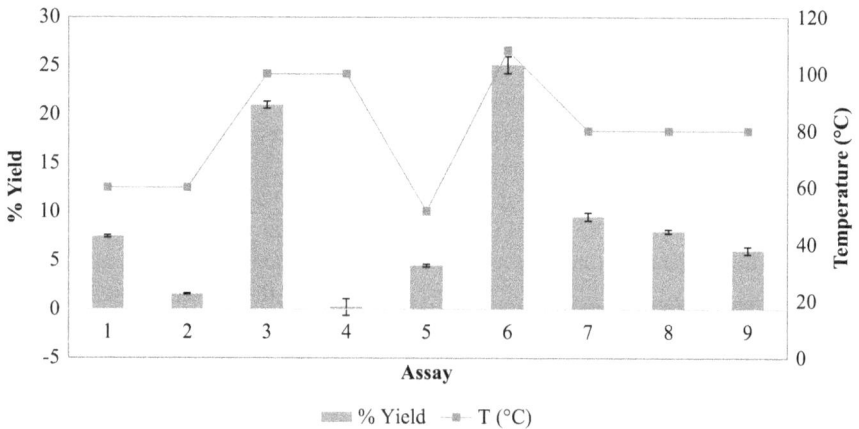

**FIGURE 5.5** Yield and temperature charts for each experiment.

*Note:* Error bars represent scatter bars.

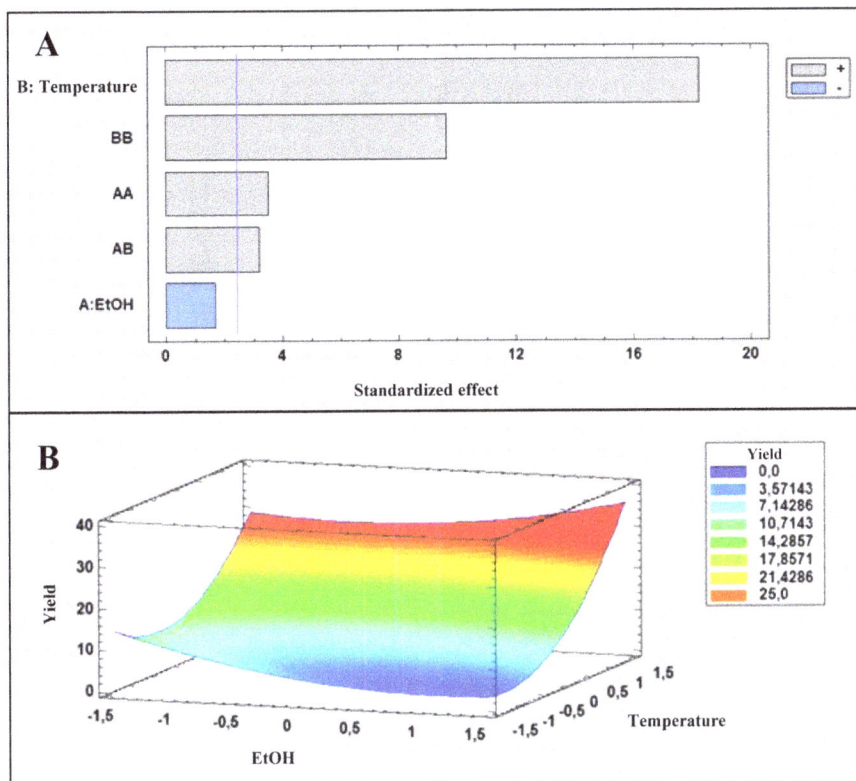

**FIGURE 5.6**  (A) Pareto diagram for performance. (B) Yield response surface.

surface graph (Figure 5.6B), whereas when temperature increases, a higher extraction yield is shown.

The highest total phenolic compound content (TPC) was 57.53 mg GAE/g extract at point 6 (108.3°C and 50% EtOH) followed by 234.96 mg GAE/g extract, while the assay displaying the lowest TPC was point 2 (60°C and 60% EtOH), 165.87 mg GAE/g extract (Table 5.2). This variable is influenced by the extraction temperature (Figure 5.7).

In the Pareto diagram (Figure 5.8A) temperature is corroborated as the factor with the most substantial influence, as shown for the response surface (Figure 5.8B) the points with the better TPC were those tested applying a higher temperature that promotes the activity of solvents and analytes, breaking complex bonds (hydrogen bonds to phenolic compounds) and thus facilitating their extraction (Muñoz et al., 2015). The stability of phenolic compounds was validated in a study by Liazid in 2007, roughly recovering 95% of TPC at 125°C and showing degradation of phenolic compounds at higher temperatures (Liazid et al., 2007). An optimal potentializing temperature for phenolic compounds was found at a 120°C cap, and avoiding exposure to temperatures over 130°C should be considered. Of note, starting at

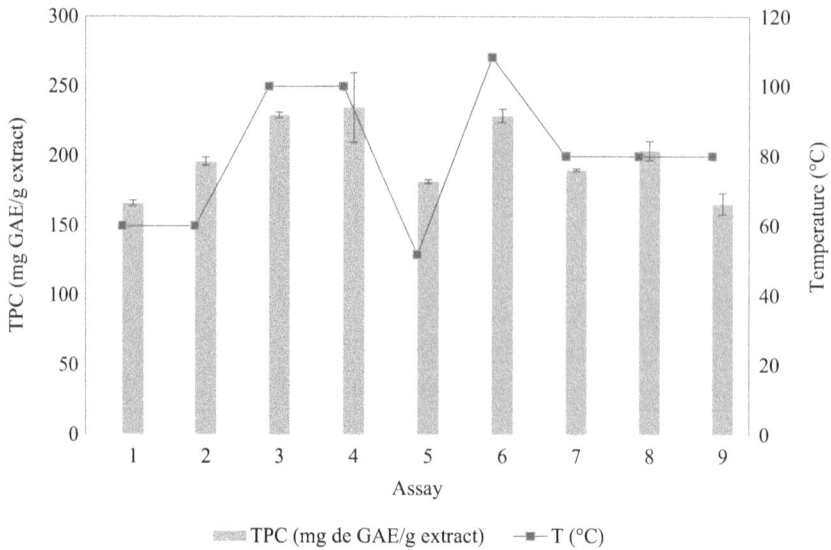

**FIGURE 5.7**    Total phenol concentration and temperature by assay plot.

*Note:* Error bars represent scatter bars.

150°C, damage to these bioactive compounds is very likely (Sharma et al., 2015; Ghafoor et al., 2019).

Regarding the total flavonoid content (TFC) the best result was obtained, as for previous cases, in study 6 (108.3°C, 50% EtOH), resulting in 756.94 µg quercetin/g extract. The lowest result was obtained in point 2 (60°C, 60% EtOH) with 590.36 µg quercetin/g extract (Figure 5.9).

Particularly, for the flavonoids variable, it can be observed in the Pareto diagram (Figure 5.10A) that the factor having the most influence on the TFC was EtOH percentage. Likewise, in the response surface graph (Figure 5.10B) the significance of temperature is observed, however, by increasing the ethanol percentage in this case temperature has a more relevant influence on collecting the TFC.

For the antioxidant activity assessed by the DPPH method, a higher value was presented in assay 4 (100°C and 60% EtOH) with a value of 5.66 mmol Trolox/g extract followed by assay 3 with a value of 5.52 mmol Trolox/g extract (100°C and 40% ethanol) (Figure 5.11).

The factor mostly influencing the antioxidant activity is temperature, as can be seen in the Pareto diagram (Figure 5.12A) and the response surface graph (Figures 5.12B). A directly proportional behavior between the variables TPC and AA is then established so the higher the amount of phenolic compounds extracted the higher the antioxidant activity observed.

Based on quadratic models where acceptability of results for each variable can be seen, obtaining a coefficient of determination $R^2$ indicates the efficiency of the model presented by the study, where the closeness of the results to the fitted regression line is shown by this measure; the higher the $R^2$ percentage as the variable values in

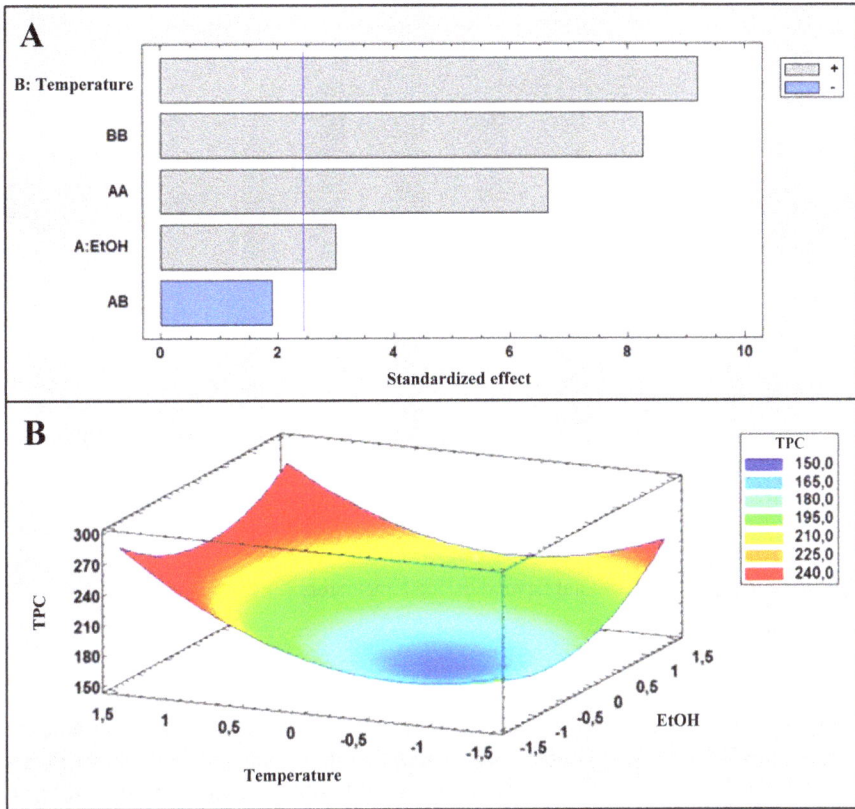

**FIGURE 5.8** (A) Pareto diagram for total phenol content. (B) Response surface for total phenol content.

the proposed model are changed the greater the amount of results lying close to the regression line, i.e., the higher $R^2$ is the higher the reliability of the model will be. In the same way, a $P$ value is obtained which is the uncertainty value to determine if the results from our model are correct (Table 5.4). Based on this, optimization and further prediction for every variable of the process was carried out at the end of the study (Table 5.5). For temperature, the optimal value is 107.87°C and for ethanol % it is 52.66%, a point not found in the experimental testing of this research. However, the optimal point closest condition between assays was point 6, where the best results were obtained.

The results shown in Table 5.5, as mentioned above, represent the optimal values for each variable, indicating the potential to use these extracts as substitutes for common industrial antioxidants. In this study, direct relationships between temperature and the amount of compounds obtained were observed since extractions conducted at a higher temperature led to better results, unlike variations of concentrations and rates where no significant changes in the characteristics of extracts obtained were exhibited.

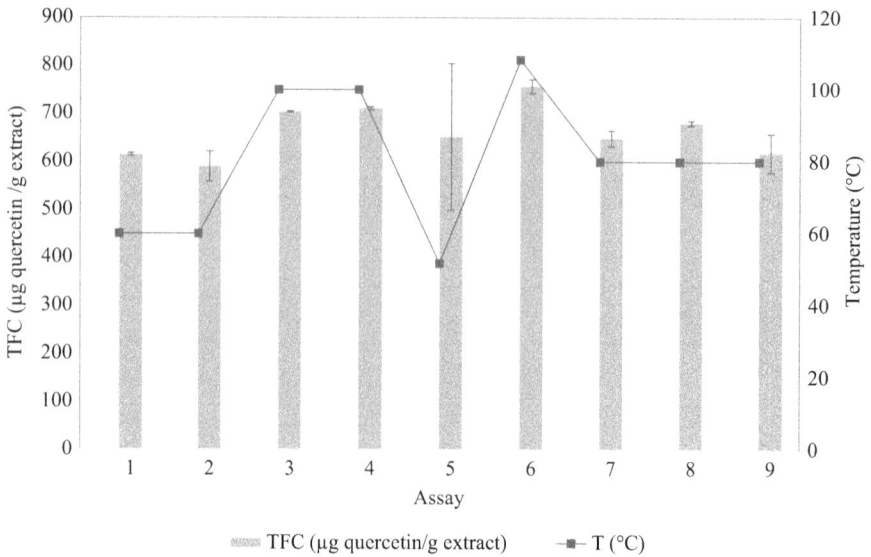

**FIGURE 5.9** (A) Graph of total flavonoids and temperature for each extract.

*Note:* Error bars represent scatter bars.

On the other hand and overall, in the studies carried out on extraction and analysis of Isabella grape properties, it was found that in the study carried out by Pereira et al. (2019), the Isabella grape pomace was used to obtain extracts by PLE using an extraction temperature of 100°C, ethanol:water (1:1) as solvent, and pH=2.0 acidified with citric acid obtaining a TPC of 65.68 mg GAE/g extract and an antioxidant activity of 7.721 mmol Trolox/g extract. The TPC reported in the study by Pereira et al. is higher as compared to the experiment from this study, with the best result being 57.53 mg GAE/g extract in point 6 (108.3°C and 50% EtOH). In addition, the results for antioxidant activity are higher than those from our study's best result experiment (assay 4; 100°C and 60% EtOH) with a value of 5.66 mmol Trolox/g extract, which could be explained by the solvent acidification that would improve the yield for phenol extraction processes, thus improving the antioxidant activity, taking into account the correlations of such compounds having this property. In another study, it was determined that the assay yielding the highest extraction on PLE was under conditions of temperature of 100°C and pressure of 12 MPa, and ethanol:water (1:1) as solvent, with 7.1% being the best extraction yield and obtaining a TPC of 231.32 ± 4.93 mg GAE/g extract as the highest result that was derived from Tintilla grapes (Otero-Pareja et al., 2015). Despite being lower in that design it was found that temperature is a pivotal parameter that influences the experimental performance of grape pomace extraction by PLE as in the present study. On the other hand, in that same study, the extraction yield was lower than that from our assays where the point with the values of higher similarity to the variables of the compared study is experiment 6 (108.3°C and 50% EtOH) being the extract that displayed the highest yield

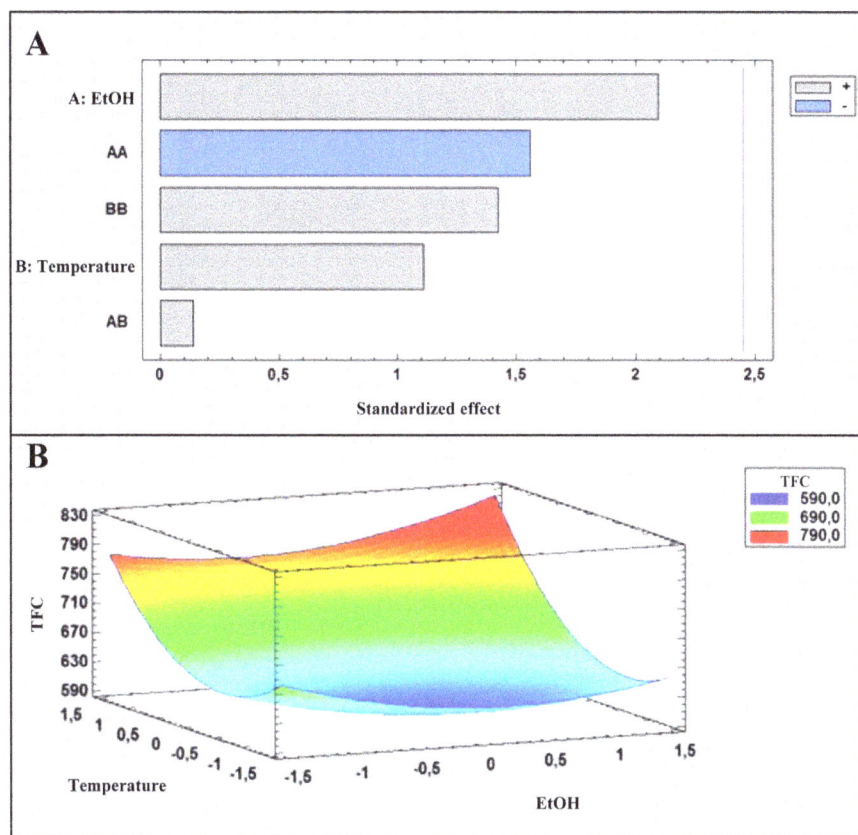

**FIGURE 5.10** (A) Pareto diagram for total flavonoid content. (B) Response surface for total flavonoid content.

(25.1%). In this same experiment, the second highest TPC result is also obtained with a value of 234.96 mg GAE/g of extract, being slightly higher than the results for phenol content obtained by Otero-Pareja et al.

Similarly, other types of biomass such as that from *Sideritis scardica* and *Clinopodium vulgare* (dried herbs) have been studied (Mihaylova et al., 2014) through PLE extraction carried out for 1.45 h using a 70% solution of ethanol:water (v/v) containing 0.1 M HCl (pH 3.6), the polyphenol total contents from *Clinopodium vulgare* and *Sideritis scardica* were 2.81 and 1.31 mg GAE/100 g of dry sample, respectively. For the antioxidant activity by DPPH radicals, values of 0.172 and 0.061 mmol Trolox/g of extract were reported for *C. vulgare* and *S. scardica*, respectively. Compared to our study, the amount of polyphenols recovered from both dry herb species in the Mihaylova study is much lower than the results from this research which cover values between 3.19 and 57.53 mg GAE/100 g of pomace. Furthermore, the antioxidant activity is lower than that observed in this study, with ranges between 1.17 and 5.66 mmol Trolox/g of extract. The dramatic difference in the obtained

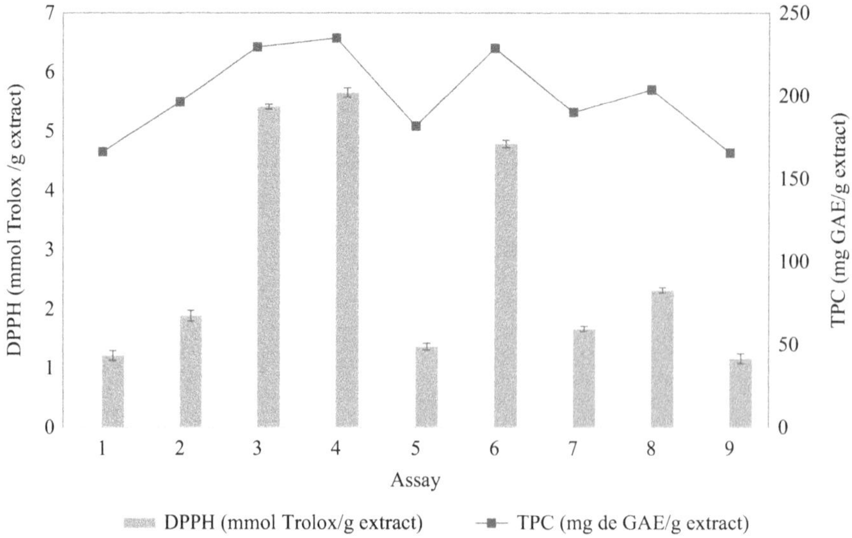

**FIGURE 5.11**　Graph of DPPH and TPC by extract.

*Note:* Error bars represent scatter bars.

results could well be given by the difference of bioactive compounds found in dried herbs and Isabella grape, with the latter being claimed as a more significant source of phenolic compounds.

In addition to the above, studies carried out on food such as that by Freitas et al. (2017) analyzed the Isabella grape seed extract obtained by agitation with ethanol, where the antioxidant activity was assessed and oxidative protection in soybean oil at 180°C for 30 minutes was compared. Contrasting the extract and the common industrial antioxidant BHT, the results obtained for the TPC were 118.39 mg GAE/g of seeds, lower than the values reported in this study and taking into account that the extraction methods were different, and that the recovery potential of PLE for the desired analyte is higher given its characteristics. However, in the study by Freitas et al. (2017), a protective effect against thermal oxidation for Isabella grape seed extract was demonstrated and similar stability between BTH soybean oil and grape extract oil was observed, with the extract–BTH mixture having a higher antioxidant capacity compared to soybean oil only added to BTH. On the other hand, other studies such as the one presented by Libera et al. (2020) looked at the efficiency of grape seed extract as a natural antioxidant and sodium ascorbate (a chemical source antioxidant) and it was compared by adding the extract and ascorbate to pork meat at maturation for 2 months. Similarly to this study, the data reported by Liberia et al. also showed that the grape seed extract had a representative amount of total phenolic compounds and antioxidant capacity, with values of 6.74 mg of GAE/ml of extract and 5.14 mmol of Trolox/ml of extract, respectively. It was also found that the grape seed extract had an efficacy close to that of sodium ascorbate for counteracting food oxidation, as expected from laboratory testing data.

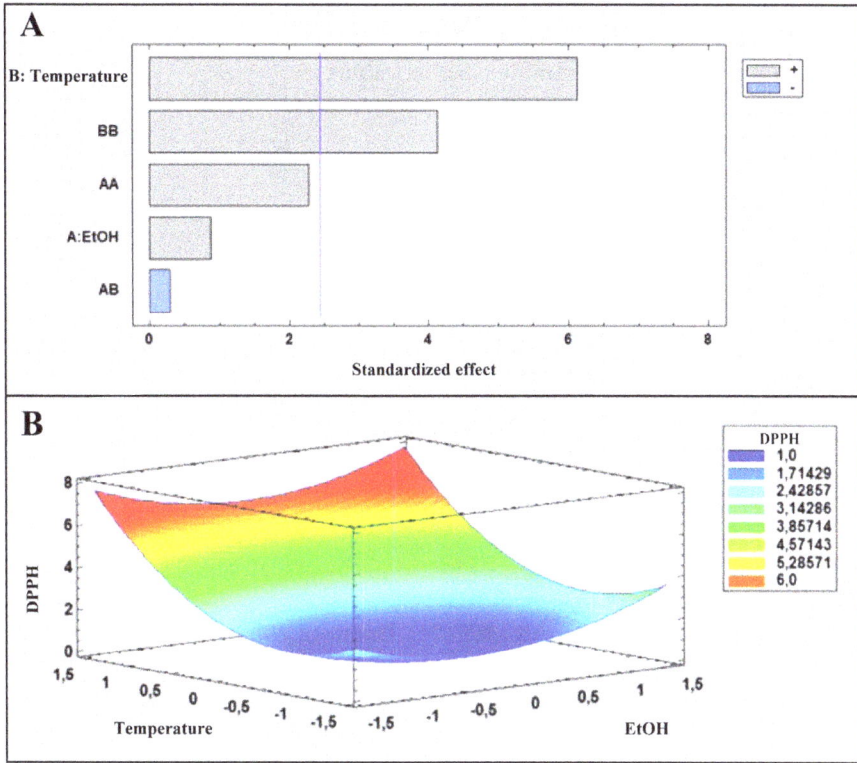

FIGURE 5.12 (A) Pareto diagram for antioxidant activity. (B) Response surface for antioxidant activity.

## TABLE 5.4
## Table for variables and quadratic models

| Variable | Quadratic Model | $R^2$ | $P$-value |
|----------|-----------------|-------|-----------|
| Yield | $6.05 - 0.74*A + 8.02*B + 1.73*A^2 + 2.00*A*B + 4.75*B^2$ | 98.66 | 0.14 |
| TPC | $165.87 + 6.86*A + 21.05*B + 16.99*A^2 - 6.15*A*B + 21.13*B^2$ | 96.96 | 0.02 |
| TFC | $618.46 + 84.47*A + 44.81*B - 70.27*A^2 + 7.89*A*B + 64.25*B^2$ | 65.12 | 0.08 |
| TEAC | $1.17 + 0.23*A + 1.60*B + 0.66*A^2 - 0.11*A*B + 1.21*B^2$ | 90.60 | 0.41 |

*Note:* A = ethanol; B = temperature; TPC = total phenol content; TFC = total flavonoid content; TEAC = Trolox-equivalent antioxidant capacity.

### TABLE 5.5
### Prediction for optimal values

| Prediction | |
| --- | --- |
| Yield | 27.4606 |
| TPC | 238.186 |
| TFC | 757.27 |
| DPPH | 5.8895 |

*Note:* Yield = % TPC = total phenol content (mg
GAE/g extract); TFC = total flavonoid con-
tent (µg quercetin /g extract); DPPH = (mmol
Trolox/g extract).

## CONCLUSIONS

By fulfilling the objective, a process for recovery of phenolic antioxidants due to extraction using pressurized liquids was developed achieving outstanding results for the collection of bioactive compounds from Isabella grapes, and corroborating their efficacy vs. other extraction methods.

Regarding optimization of extraction-associated variables based on the recovery of phenolic compounds and antioxidant properties, 107.87°C and ETOH at 52.66% were found to be the optimal point. Temperature was reported as the process param-eter with the major effect on every response variable analyzed. On the other hand, since the optimal point is not within the tests performed, the closest value was taken, which was also the one displaying the best yield and total phenol content at 108.3°C and 50% EtOH.

Regarding the antioxidant activity, a direct proportional relationship with the amount of total phenol obtained from extracts was found. The point of 5.66 mmol Trolox per gram of extract, which demonstrates a higher antioxidant capacity, was the testing point where the highest amount of total phenol was recovered (54.31 mg GAE per 100 grams).Therefore, it can be stated that PLE is an effective technique for the extraction of bioactive compounds with potential antioxidant activity from agro-industrial wine-making residues. This could lead to the use of these extracts as additives in different industries to replace synthetic compounds that are harmful for health, while also reducing the untoward effects on the environment.

## GLOSSARY

**Agro-industrial waste:** Biological mass from agro-industrial sources which does not have a specific disposition, in most cases it is left in the field for natural decomposition (Simbaña, 2010).

**Analyte:** A sample component with analytical interest. They are chemical species whose presence or concentration is desired to be known. Analytes are identifiable and quantifiable (Mencias, 2019).

**Anthocyanins:** Phenolic pigments soluble in water, specifically belonging to the flavonoids group, with antioxidant properties.Anthocyanins are glycosides with a sugar in position 3 and also at some other position, and are responsible for most of the colors in red, purple, and blue tones of plants, fruits, or vegetables (Khoo et al., 2017).

**Antioxidants:** Compounds capable of slowing down, delaying, or preventing the oxidation of an oxidizable material (Amorati et al., 2013).

**Bioactive compound:** Compounds mostly obtained from plants and characterized by having important beneficial properties for human health, with flavonoids being the group encompassing the majority of bioactive compounds (Martínez-Navarrete et al. (2008).

**Biomass:** Biodegradable fraction from agriculture, forestry, and/or agro-industry or municipal and industrial waste products, waste and residues that can be converted into other products (García Morales et al., 2014).

**Biorefinery:** First defined in Germany in 1997. According to this definition, the "green biorefineries" are complex systems based on green technology for the use of natural materials and energy from renewable resources. Processes to produce energy and chemicals from biomass are integrated in a biorefinery. This concept is analogous to petroleum refineries, where fuels and diverse products are obtained from oil (Schieb et al., 2015).

**Biotechnology:** Application of scientific or engineering principles to processes where a useful product is obtained from biological agents for certain material (Negrín et al., 2007).

**BPR:** Back pressure regulator valve dynamically controlled by the system outlet pressure. Without a BPR, the liquid that might be in liquid phase would expand to a low-pressure, low-density gas that has a low solvation potential (Berger, 1995).

**DPPH method:** The α-diphenyl-β-picrylhydrazyl (DPPH) is a method for removal of free radicals used to assess the antioxidant potential in a compound or extract. The method is carried out by mixing the compound or extract with the DPPH solution and absorbance is recorded after a defined period of time (Kedare & Singh, 2011).

**Extraction:** Matter transfer unitary operation consisting on the obtention of different compounds by dissolving one or more components of a mixture (either a liquid or solid sample) into a selective solvent (Costa López, 2004).

**Flavonoids:** Natural pigments in plants, vegetables, and/or fruits characterized by protecting an organism from the damage caused by oxidizing agents such as ultraviolet light or environmental pollution. Flavonoids are not produced by humans; they must be obtained by food intake or supplement consumption (Martínez-Flórez et al., 2002).

**Folin-Ciocalteau reagent:** A mixture of phosphomolybdate and phosphotungstate used in the phenolic antioxidants colorimetric assay (Chiellini, 2008).

**Free radicals:** Any chemical species, whether charged or not, having an odd electron number in the external orbital of the atomic structure and that turns out to be very unstable. Free radicals are characterized by a biradical structure, high reactivity, and a short half-life (Venereo Gutiérrez, 2002).

**Grape pomace:** Also called wine marc, this is the final residue from the winemaking process and is typically composed of pressed grapes, stem pieces, seeds, pulp, peels, and yeast cells from the wine fermentation process. Grape pomace constitutes approximately 20% of the weight of grapes used in the process (García & González, 2017).

**Phenols:** Aromatic organic compounds from the same functional group as alcohols and with some similar properties. In phenols, the C–OH bond is very difficult to break, therefore, they do not behave as bases because in phenols the OH group is directly linked to the aromatic ring exhibiting a special chemical behavior (Alliger et al., 2006).

**PLE:** Extraction performed by applying pressurized liquids. PLE is an extraction method carried out at high pressures, which favor the role of liquid solvents at high temperatures above their boiling point (Fernández, 2009).

**Spectrophotometer:** Laboratory instrument that measures the absorbance of light by molecules, so analyzing the amount and type of compound in a test sample is possible.

**Yield:** This is the amount of product obtained during the process once a reaction is carried out and can be expressed in terms of percentage based on a theoretical yield of 100% (Picado & Álvarez, 2008).

## ACKNOWLEDGMENTS

We thank the Universidad Nacional de Colombia Food Chemistry Research Group (GiQA, for the acronym in Spanish) for providing support and the facilities to make the development of this project possible. Finally, we thank the Universidad Libre for funding this project.

## REFERENCES

Agronet. (2018). Reporte: Área, Producción y Rendimiento Nacional por Cultivo. Ministerio de agricultura y desarrollo rural de Colombia. Retrieved 14 March 2023 from https://www.agronet.gov.co/estadistica/Paginas/home.aspx?cod=1"

Alliger, N. L., Cava, M. P., Johnson, C. R., Lebel, N. A., & Stevers, C. L. (1971). Química Orgánica (Segunda Ed).

Alvarez-Rivera, G., Bueno, M., Ballesteros-Vivas, D., Mendiola, J. A., & Ibáñez, E. (2019). Pressurized liquid extraction. In Liquid-Phase Extraction (pp. 375–398). Elsevier. https://doi.org/10.1016/B978-0-12-816911-7.00013-X

Amorati, R., Foti, M. C., & Valgimigli, L. (2013). Antioxidant activity of essential oils. Journal of Agricultural and Food Chemistry, 61(46), 10835–10847. https://doi.org/10.1021/jf403496k

Antoniolli, A., Fontana, A. R., Piccoli, P., & Bottini, R. (2015). Characterization of polyphenols and evaluation of antioxidant capacity in grape pomace of the cv. Malbec. Food Chemistry, 178, 172–178. https://doi.org/10.1016/j.foodchem.2015.01.082

Ballesteros, D. (2015). Estudio comparativo sobre la obtención de extractos con actividad citotóxica a partir de residuos frutícolas. Tesis de Maestria. 135. http://bdigital.unal.edu.co/52021/1/diegoballesterosvivas.2015.pdf%0A www.bdigital.unal.edu.co/52021/

Becerra, J., Lili, D., & Noreidis, P. (2017). Extracción con Líquidos Presurizados. https://es.scr ibd.com/document/363871349/Extraccion-Con-Liquidos-Presurizados-2

Berger, T. . (1995). Packed Column SFC (RSC Chromatography Monographs) (Segunda Ed). https://books.google.com.co/books?id=JzetfPqHNUQC&dq=bpr+pump&source=gbs_navlinks_s

Carabias-Martínez, R., Rodríguez-Gonzalo, E., Revilla-Ruiz, P., & Hernández-Méndez, J. (2005). Pressurized liquid extraction in the analysis of food and biological samples. Journal of Chromatography A, 1089, Issues 1–2, pp. 1–17. Elsevier. https://doi.org/10.1016/j.chroma.2005.06.072

Carr, A. G., Mammucari, R., & Foster, N. R. (2011). A review of subcritical water as a solvent and its utilisation for the processing of hydrophobic organic compounds. Chemical Engineering Journal, 172, Issue 1, pp. 1–17. Elsevier. https://doi.org/10.1016/j.cej.2011.06.007

Chandrasekaran, M. (2012). Valorization of Food Processing By-Products (Primera Ed). https://books.google.com.co/books?id=l0HNBQAAQBAJ

Chang, C.-C., Yang, M.-H., Wen, H.-M., & Chern, J.-C. (2020). Estimation of total flavonoid content in propolis by two complementary colometric methods. Journal of Food and Drug Analysis, 10(3), 178–182. https://doi.org/10.38212/2224-6614.2748

Chemat, F., & Strube, J. (2015). Green Extraction of Natural Products: Theory and Practice.

Chiellini, E. (2008). Environmentally Compatible Food Packaging. https://books.google.com.co/books?id=tk6kAgAAQBAJ&printsec=frontcover&hl=es&source=gbs_ge_summary_r&cad=0#v=onepage&q&f=false

Christen, P., & Kaufmann, B. (2014). New trends in the extraction of natural products: Microwave-assisted extraction and pressurized liquid extraction. Encyclopedia of Analytical Chemistry, 1–27. https://doi.org/10.1002/9780470027318.a9904

Costa López, J. (2004). Curso de Ingeniería Química (Primera Ed). https://books.google.com.co/books?id=XZNYpvnO_V8C&printsec=frontcover&hl=es&source=gbs_ge_summary_r&cad=0#v=onepage&q&f=false

Creasy, G., & Creasy, L. (2009). Grapes. https://books.google.com.co/books?id=pbWMPnVxN1oC&printsec=frontcover&hl=es&source=gbs_ge_summary_r&cad=0#v=onepage&q&f=false

Dávila, I., Robles, E., Egüés, I., Labidi, J., & Gullón, P. (2017). The Biorefinery Concept for the Industrial Valorization of Grape Processing By-Products. En Handbook of Grape Processing By-Products. Elsevier. https://doi.org/10.1016/B978-0-12-809870-7.00002-8

Do, L., Lundstedt, S., & Haglund, P. (2013). Optimization of selective pressurized liquid extraction for extraction and in-cell clean-up of PCDD/Fs in soils and sediments. Chemosphere, 90(9), 2414–2419. https://doi.org/10.1016/j.chemosphere.2012.10.070

Dunford, N. T., Irmak, S., & Jonnala, R. (2010). Pressurised solvent extraction of policosanol from wheat straw, germ and bran. Food Chemistry, 119(3), 1246–1249. https://doi.org/10.1016/j.foodchem.2009.07.039

Fernández, M. (2009). Estudio del comportamiento fotoquímico y determinación de compuestos fitosanitarios en matrices medioambientales y agroalimentarias mediante técnicas avanzadas de extracción y microextracción. https://books.google.com.co/books?id=o6xX6NDu3i8C&printsec=frontcover&hl=es&source=gbs_ge_summary_r&cad=0#v=onepage&q&f=false

Freitas, I. R., Cattelan, M. G., Rodrigues, M. L., Luzia, D. M. M., & Jorge, N. (2017). Effect of grape seed extract (Vitis labrusca L.) on soybean oil under thermal oxidation. Nutrition & Food Science, 47(5). https://doi.org/10.1108/NFS-04-2016-0050

García-Lomillo, J., & González-SanJosé, M. L. (2017). Applications of wine pomace in the food industry: Approaches and functions. Comprehensive Reviews in Food Science and Food Safety, 16(1), 3–22. https://doi.org/10.1111/1541-4337.12238

García Morales, J. L., Álvarez Gallego, C. josé, Paredes Gil, C., López Mosquera, E., Fernandez Morales, F. J., Bustamante Muñoz, M. Á., Barrena Gómez, R., & Seoane Labandeira, S. (2014). De Residuo a Recurso, el Camino hacia la Sostenibilidad. https://books.goo gle.com.co/books?id=XL7-CAAAQBAJ&pg=PA57&dq=residuos+subproductos&hl= es&sa=X&ved=2ahUKEwjQn86q_M3qAhUxn-AKHf53AoEQuwUwAXoECAY QBw#v=onepage&q=residuos subproductos&f=false

Gaviria Montoya, C., Hernández Arredondo, J., Lobo Arias, M., Medina Cano, C., & Rojano, B. (2012). Changes in the antioxidant activity in Mortiño fruits (*Vaccinium meridionale* Sw.) during development and ripening. Revista Facultad Nacional de Agronomía, Medellín, 65(1), 87–6495.

Ghafoor, K., Ahmed, I. A. M., Doğu, S., Uslu, N., Fadimu, G. J., Al Juhaimi, F., Babiker, E. E., & Özcan, M. M. (2019). The effect of heating temperature on total phenolic content, antioxidant activity, and phenolic compounds of plum and mahaleb fruits. International Journal of Food Engineering, 15(11–12). https://doi.org/10.1515/ijfe-2017-0302

Han, X., Ding, C., He, L., Hu, Q., & Yu, S. (2011). Development of loop-mediated isothermal amplification (LAMP) targeting the GroEL gene for rapid detection of *Riemerella anatipestifer*. Avian Diseases Digest, 6(3), e33–e34. https://doi.org/10.1637/9795-960 211-digest.1

Herrero, M., Castro-Puyana, M., Mendiola, J. A., & Ibañez, E. (2013). Compressed fluids for the extraction of bioactive compounds. In TrAC – Trends in Analytical Chemistry, 43, 67–83. Elsevier B.V. https://doi.org/10.1016/j.trac.2012.12.008

Hid Cadena, R., Bautista Ortín, A. B., Ortega Regules, A. E., Welti Chanes, J. S., Lozada Ramírez, J. D., & Anaya de Parrodi, C. (2010). Cambios en contenido de compuestos fenólicos y color de extractos de Jamaica (Hibiscus sabdariffa) sometidos a calentamiento con energía de microondas. IX Congreso Nacional Del Color: Alicante, 29 y 30 de Junio, 1 y 2 de Julio de 2010, 1(October 2015), 303. http://hdl.handle.net/10045/16463

Howard, L., & Pandjaitan, N. (2008). Pressurized liquid extraction of flavonoids from spinach. Journal of Food Science, 73(3). https://doi.org/10.1111/j.1750-3841.2007.00658.x

International Organisation of Vine and Wine. (2022). Database. Retrieved 14 March 2023 from https://www.oiv.int/what-we-do/data-discovery-report?oiv"

Katalinić, V., Možina, S. S., Skroza, D., Generalić, I., Abramovič, H., Miloš, M., Ljubenkov, I., Piskernik, S., Pezo, I., Terpinc, P., & Boban, M. (2010). Polyphenolic profile, antioxidant properties and antimicrobial activity of grape skin extracts of 14 Vitis vinifera varieties grown in Dalmatia (Croatia). Food Chemistry, 119(2), 715–723. https://doi.org/10.1016/j.foodchem.2009.07.019

Kedare, S. B., & Singh, R. P. (2011). Genesis and development of DPPH method of antioxidant assay. Journal of Food Science and Technology, 48(4), 412–422. https://doi.org/10.1007/s13197-011-0251-1

Khoo, H. E., Azlan, A., Tang, S. T., & Lim, S. M. (2017). Anthocyanidins and anthocyanins: Colored pigments as food, pharmaceutical ingredients, and the potential health benefits. Food and Nutrition Research, *61*. https://doi.org/10.1080/16546 628.2017.1361779

Kultys, E., & Kurek, M. A. (2022). Green extraction of carotenoids from fruit and vegetable byproducts: A review. Molecules, 27(2). https://doi.org/10.3390/molecules

Liazid, A., Palma, M., Brigui, J., & Barroso, C. G. (2007). Investigation on phenolic compounds stability during microwave-assisted extraction. Journal of Chromatography A, 1140(1–2). https://doi.org/10.1016/j.chroma.2006.11.040

Libera, J., Latoch, A., & Wójciak, K. M. (2020). Utilization of grape seed extract as a natural antioxidant in the technology of meat products inoculated with a probiotic strain of LAB. Foods, 9(1). https://doi.org/10.3390/foods9010103

Lucarini, M., Durazzo, A., Romani, A., Campo, M., Lombardi-Boccia, G., & Cecchini, F. (2018). Bio-based compounds from grape seeds: A biorefinery approach. Molecules, 23(8), 1–12. https://doi.org/10.3390/molecules23081888

Magalhães, L. M., Barreiros, L., Maia, M. A., Reis, S., & Segundo, M. A. (2012). Rapid assessment of endpoint antioxidant capacity of red wines through microchemical methods using a kinetic matching approach. Talanta, 97, 473–483. https://doi.org/10.1016/j.talanta.2012.05.002

Makris, D. P., Kallithraka, S., & Kefalas, P. (2006). Flavonols in grapes, grape products and wines: Burden, profile and influential parameters. Journal of Food Composition and Analysis, 19(5), 396–404. Academic Press. https://doi.org/10.1016/j.jfca.2005.10.003

Martínez-Flórez, S., González-Gallego, J., Culebras, J. M., & Tuñón, M. J. (2002). Flavonoids: properties and anti-oxidizing action. Nutricion Hospitalaria, 17(6), 271–278. www.ncbi.nlm.nih.gov/pubmed/12514919

Martínez-Navarrete, N., del Mar Camacho Vidal, M., & José Martínez Lahuerta, J. (2008). Los compuestos bioactivos de las frutas y sus efectos en la salud. Actividad Dietética, 12(2), 64–68. doi:10.1016/s1138-0322(08)75623-2

Mejías, N., Orozco, E., & Galáan, H. (2016). Aprovechamiento de los residuos agroindustriales y su contribución al desarrollo sostenible de México Revista de Ciencias Ambientales y Recursos Naturales. Revista de Ciencias Ambientales y Recursos Naturales, 2(6), 27–41.

Mencias, J. (2019). Resonancia Magnética Nuclear. https://books.google.com.co/books?id=3FWIDwAAQBAJ&printsec=frontcover&hl=es&source=gbs_ge_summary_r&cad=0#v=onepage&q&f=false

Mihaylova, D., Lante, A., & Krastanov, A. I. (2014). A study on the antioxidant and anti-microbial activities of pressurized-liquid extracts of Clinopodium vulgare and Sideritis scardica. Agro FOOD Industry Hi Tech-Vol, 25(6), 55–58.

Miron, T. L., Plaza, M., Bahrim, G., Ibáñez, E., & Herrero, M. (2011). Chemical composition of bioactive pressurized extracts of Romanian aromatic plants. Journal of Chromatography A, 1218(30), 4918–4927. https://doi.org/10.1016/j.chroma.2010.11.055

Muñoz, C., Chavez, R., Ludy, C., Rendón, F., Margarita, R., Ángela, M., Pabón, & Otálvaro-álvarez, Á. (2015). Extracción de compuestos fenólicos con actividad antioxidante a partir de Champa (Campomanesia lineatifolia). Revista CENIC. Ciencias Químicas, 46, 38–46. www.redalyc.org/articulo.oa?id=181643224027

Mustafa, A., & Turner, C. (2011). Pressurized liquid extraction as a green approach in food and herbal plants extraction: A review. Analytica Chimica Acta, 703(1), 8–18. https://doi.org/10.1016/j.aca.2011.07.018

Negrín, S., Sosa, Á. E., Ayala, M., Diosdado, E., Pérez, M. R., Pujol, M., Fernández, J. R., Muzio, V., Castellanos, L., González, L. J., Cremata, J., Quintana, M., Pérez, G., Valdés, J., Rodríguez, M. P., Borroto, C., González, C., Morales, J., Duarte, C., … Lage, A. (2007). Enseñanza popular de la Biotecnología. 24(1), 54–58. https://biblat.unam.mx/hevila/Biotecnologiaaplicada/2007/vol24/no1/14.pdf

Otero-Pareja, M. J., Casas, L., Fernández-Ponce, M. T., Mantell, C., & De La Ossa, E. J. M. (2015). Green extraction of antioxidants from different varieties of red grape pomace. Molecules, 20(6), 9686–9702. https://doi.org/10.3390/molecules20069686

Panja, P. (2018). Green extraction methods of food polyphenols from vegetable materials. Current Opinion in Food Science, 23, 173–182. https://doi.org/10.1016/J.COFS.2017.11.012

Pasrija, D., & Anandharamakrishnan, C. (2015). Techniques for extraction of green tea polyphenols: A review. Food and Bioprocess Technology, 8(5), 935–950. https://doi.org/10.1007/s11947-015-1479-y

Pereira, D. T. V., Tarone, A. G., Cazarin, C. B. B., Barbero, G. F., & Martínez, J. (2019). Pressurized liquid extraction of bioactive compounds from grape marc. Journal of Food Engineering, 240. https://doi.org/10.1016/j.jfoodeng.2018.07.019

Picado, M. B., & Álvarez, M. (2008). Química I Introducción al Estudio de la Materia (Primera Ed). https://books.google.com.co/books?id=mjvKG4BJ0xwC&printsec=frontcover&hl=es&source=gbs_ge_summary_r&cad=0#v=onepage&q&f=false

Picó, Y. (2017). Pressurized liquid extraction of organic contaminants in environmental and food samples. Comprehensive Analytical Chemistry, 76, 83–110. https://doi.org/10.1016/bs.coac.2017.03.004

Plaza, M., & Turner, C. (2015). Pressurized hot water extraction of bioactives. TrAC – Trends in Analytical Chemistry, 71, 39–54). https://doi.org/10.1016/j.trac.2015.02.022

Poole, C. F. (2020). Milestones in the development of liquid-phase extraction techniques. Liquid-Phase Extraction, 1–44). https://doi.org/10.1016/B978-0-12-816911-7.00001-3

Prieto, R., Gonzalez, G., & Abella, J. (2011). Extracción con ultrasonido de compuestos fenólicos presentes en la cáscara de uva variedad Isabella (Vitis labrusca). Reciteia, 11(1b), 117–126."

Richter, B. E., & Raynie, D. (2012). Accelerated Solvent Extraction (ASE) and High-Temperature Water Extraction. In Comprehensive Sampling and Sample Preparation. Elsevier. https://doi.org/10.1016/B978-0-12-381373-2.00047-8

Santos, D., Gomes, M. T. M., Vardanega, R., & Rostagno, M. (2013). Integration of Pressurized Fluid-based Technologies for Natural Product Processing. In M. Rostagno & J. M. Prado (Eds.), Natural Product Extraction: Principles and Applications (pp. 399–441). Royal Society of Chemistry.

Sharma, K., Ko, E. Y., Assefa, A. D., Ha, S., Nile, S. H., Lee, E. T., & Park, S. W. (2015). Temperature-dependent studies on the total phenolics, flavonoids, antioxidant activities, and sugar content in six onion varieties. Journal of Food and Drug Analysis, 23(2). https://doi.org/10.1016/j.jfda.2014.10.005

Schieb, P.-A., Lescieux-Katir, H., Thénot, M., & Clément-Larosière, B. (2015). Biorefinery 2030. Springer: Berlin, Heidelberg. https://doi.org/10.1007/978-3-662-47374-0

Segura, C., Guerrero, C., Posada, E., Mojica, J., & Pérez, W. (2015). Caracterización de residuos de la industria vinícola del valle de Sáchica con potencial nutricional para su aprovechamiento después del proceso agroindustrial. Encuentro Nacional de Investigación y Desarrollo (ENID), 1–16. https://doi.org/10.13140/RG.2.1.3024.8406

Simbaña, A. (2010). Fibras naturales y residuos agroindustriales. Fuente sotenible de materia prima. Revista Científica Axioma, 1(6), 15–16. http://pucesinews.pucesi.edu.ec/index.php/axioma/article/view/308/299

Singleton, V. L., Orthofer, R., & Lamuela-Raventós, R. M. (1999). Analysis of total phenols and other oxidation substrates and antioxidants by means of folin-ciocalteu reagent. https://doi.org/10.1016/S0076-6879(99)99017-1

Tamires Vitor Pereira, D., Vollet Marson, G., Fernández Barbero, G., Gadioli Tarone, A., Baú Betim Cazarin, C., Dupas Hubinger, M., & Martínez, J. (2020). Concentration of bioactive compounds from grape marc using pressurized liquid extraction followed by integrated membrane processes. Separation and Purification Technology, 250. https://doi.org/10.1016/j.seppur.2020.117206

Turner, C., & Waldebäck, M. (2013). Principles of pressurized fluid extraction and environmental, food and agricultural applications. In: Separation, Extraction and Concentration

Processes in the Food, Beverage and Nutraceutical Industries. Elsevier. https://doi.org/10.1533/9780857090751.1.67

Vargas y Vargas, M. de L., Figueroa Brito, H., Tamayo Cortez, J. A., Toledo López, V. M., & Moo Huchin, V. M. (2019). Aprovechamiento de cáscaras de frutas: análisis nutricional y compuestos bioactivos. CIENCIA Ergo Sum, 26(2). https://doi.org/10.30878/ces.v26n2a6

Venereo Gutiérrez, J. (2002). Daño oxidativo, radicales libres y antioxidantes. Rev Cub Med Mil, 31(2).

Waterhouse, A. L. (2003). Determination of Total Phenolics. In: Current Protocols in Food Analytical Chemistry. John Wiley & Sons, Inc. https://doi.org/10.1002/0471142913.fai0101s06

Wianowska, D., & Gil, M. (2019). Critical approach to PLE technique application in the analysis of secondary metabolites in plants. TrAC Trends in Analytical Chemistry, 114. https://doi.org/10.1016/j.trac.2019.03.018

Woisky, R. G., & Salatino, A. (1998). Analysis of propolis: some parameters and procedures for chemical quality control. Journal of Apicultural Research, 37(2). https://doi.org/10.1080/00218839.1998.11100961

Yang, Y., Bowadt, S., Hawthorne, S. B., & Miller, D. J. (1995). Subcritical water extraction of polychlorinated biphenyls from soil and sediment. Analytical Chemistry, 67(24). https://doi.org/10.1021/ac00120a022

Yepes, S. M., Montoya, L. J., & Orozco, F. (2008). Valorización de residuos agroindustriales – frutas – en Medellín y el sur del valle del aburrá, Colombia. Rev. Fac. Nal. Agr., 61(1).

## ANNEXES

### ANNEX 5.1
### Table of extraction conditions and yield

| Assay | D (%EtOH) | Sample (g) | Extract (g) | Time (h) | % Yield |
|---|---|---|---|---|---|
| 1 | 40.0% | 4.5950 | 0.3407 | 2 | 7.50 |
| 2 | 60.0% | 4.7046 | 0.0783 | 2 | 1.60 |
| 3 | 40.0% | 4.9006 | 1.0429 | 2 | 21.00 |
| 4 | 60.0% | 4.8397 | 1.1479 | 2 | 23.10 |
| 5 | 50.0% | 4.8799 | 0.2244 | 2 | 4.50 |
| 6 | 50.0% | 4.7154 | 1.1569 | 2 | 25.10 |
| 7 | 35.9% | 4.8458 | 0.4464 | 2 | 9.50 |
| 8 | 64.1% | 5.4605 | 0.4308 | 2 | 8.00 |
| 9 | 50.0% | 4.9568 | 0.2932 | 2 | 6.20 |
| 10 | 50.0% | 5.0223 | 0.3438 | 2 | 6.70 |
| 11 | 50.0% | 4.7541 | 0.2395 | 2 | 5.10 |
| 12 | 50.0% | 5.2558 | 0.3295 | 2 | 6.20 |

## ANNEX 5.2
### Table of the study overall results

| Assay | D (%EtOH) | T (°C) | Yield | Total phenols mg /g extract | | | | | Total phenols mg de GAE/ 100 g | Total flavonoid µg QE /g extract | | | | | DPPH mmol Trolox /g extract | | | | |
|---|---|---|---|---|---|---|---|---|---|---|---|---|---|---|---|---|---|---|---|
| | | | | r1 | r2 | Average | Standard deviation | CV | Average | r1 | r2 | Average | Standard deviation | CV | r1 | r2 | Average | Standard deviation | CV |
| 1 | 40.0% | 60.0 | 7.5% | 167.61 | 164.91 | 166.26 | 1.90 | 1.15 | 12.47 | 612.56 | 617.19 | 614.87 | 3.27 | 0.53 | 1.16 | 1.28 | 1.22 | 0.08 | 6.76 |
| 2 | 60.0% | 60.0 | 1.6% | 194.14 | 198.44 | 196.29 | 3.04 | 1.55 | 3.19 | 568.19 | 612.52 | 590.36 | 31.35 | 5.31 | 1.95 | 1.83 | 1.89 | 0.09 | 4.65 |
| 3 | 40.0% | 100.0 | 21.0% | 230.89 | 228.20 | 229.54 | 1.90 | 0.83 | 48.25 | 705.12 | 702.81 | 703.97 | 1.64 | 0.23 | 5.45 | 5.39 | 5.42 | 0.04 | 0.76 |
| 4 | 60.0% | 100.0 | 23.1% | 252.66 | 217.25 | 234.96 | 25.04 | 10.66 | 54.31 | 713.38 | 708.70 | 711.04 | 3.31 | 0.47 | 5.72 | 5.60 | 5.66 | 0.08 | 1.47 |
| 5 | 50.0% | 51.7 | 4.5% | 182.82 | 180.94 | 181.88 | 1.33 | 0.73 | 8.18 | 759.74 | 543.79 | 651.76 | 152.70 | 23.43 | 1.33 | 1.42 | 1.37 | 0.06 | 4.18 |
| 6 | 50.0% | 108.3 | 25.1% | 232.23 | 225.46 | 228.85 | 4.79 | 2.09 | 57.53 | 746.47 | 767.42 | 756.94 | 14.81 | 1.96 | 4.75 | 4.83 | 4.79 | 0.06 | 1.30 |
| 7 | 35.9% | 80.0 | 9.5% | 190.89 | 189.53 | 190.21 | 0.96 | 0.51 | 18.07 | 660.22 | 636.87 | 648.54 | 16.51 | 2.55 | 1.70 | 1.64 | 1.67 | 0.04 | 2.49 |
| 8 | 64.1% | 80.0 | 8.0% | 199.15 | 208.80 | 203.97 | 6.82 | 3.34 | 16.38 | 683.94 | 676.83 | 680.38 | 5.03 | 0.74 | 2.35 | 2.29 | 2.32 | 0.04 | 1.82 |
| 9 | 50.0% | 80.0 | 6.05% | 160.46 | 171.29 | 165.87 | 7.66 | 4.62 | 10.03 | 590.53 | 646.39 | 618.46 | 39.50 | 6.39 | 1.23 | 1.11 | 1.17 | 0.08 | 7.10 |

# 6 Evaluation of Anti-quorum Sensing and Antibacterial Activity in Isabella Grape (*Vitis labrusca* L.) Pomace Extracts

*Daniela Méndez Velásquez, Faride Geraldine Jiménez Rodríguez, and Patricia Joyce Pamela Zorro Mateus*

Agri-business is defined as the industrial economic activity for the development of food products or raw materials intended for human production and consumption (González, 2017, pp. 142–143). The sector is currently considered by the FAO (Food Agriculture Organization) to be one of the largest waste generators worldwide, where up to approximately one-third of food for human consumption is discarded, including production, management, processing, and marketing processes (Cury et al., 2017, pp. 122–128).

According to previous studies, around 2010 million tons/year of solid waste were estimated to be generated by large cities globally by 2018, with a per capita production of approximately 4995.73 kg/day, thus it is projected that by 2050 this figure will increase globally to 3400 million tons/year with values per capita of 8450.48 kg/day (Silpa et al., 2018, pp. 24–25). In addition, it was estimated that in Colombia about 12 million tons of solid waste are generated per year, from which only 17% is used (Semana magazine, 2020), thus increasing the wastage of raw materials as a potential source of valorization to generate by-products.

The rice, chocolate, coffee, cattle and pork food, sugar, wine, potato, banana, and cassava industries represent the agro-industrial sector in Colombia, as well as the fruit (mostly) and vegetable sector, oil and fat factories, rubber derivatives, and dairy products. These industries mostly use agricultural and fruit products as raw materials and essential inputs in their production processes (Cury et al., 2017, p. 123). It should be noted that according to the National Agricultural Survey (ENA, for the acronym in Spanish to Encuesta Nacional Agropecuaria), the production from fruit and

DOI: 10.1201/9781003391593-6

agricultural industries in the country was 63,247,863 tons for the year 2019, of which 42,208,363 tons corresponded to the agro-industrial group (66.7%), and 6,712,167 tons to fruit crops (10.6%) (National Administrative Statistics Department [DANE], 2020). Based on estimates by the National Planning Department (DNP) in Colombia, for 2016, about 34% of food waste is lost at the agricultural step, the post-harvest and storage process, the industrial process, and at distribution and retail. Given the above, these wastes in most cases are incinerated or taken to sanitary landfills (Beleño, 2018; Chávez & Rodríguez, 2016, p. 92; Food and Agriculture Organization, [FAO], 2017), which has major environmental implications.

In Colombia, the wine-making industry derived from the fruit industry has been steadily increasing both at the national and international levels, and is characterized by using Isabella grapes as the raw material in the production processes, from which the grape pomace is derived as a by-product. In Colombia, grape production was estimated to be around 30,000 tons per year, where 85% of the total national production for this crop is in the Valle del Cauca department representing 86.9% of the total area sown throughout the country (Pinilla, 2016). This massive yearly grape tonnage favors a considerable increase in agro-industrial waste represented as peels, seeds, and leachates obtained from pressing of the fruit.

Organic waste or by-products derived from the wine-making industry are usually taken to sanitary landfills, causing significant ecological damage and representing significant losses to essential natural resources.

Specifically, grape pomace is an important source of sugars, cellulose, lignins, lipids, and phenolic compounds. Some of these compounds are environmental pollutants or their precursors, as is the case of lipids and carbohydrates, which require large quantities of oxygen for degradation and highly expensive environmental handling. Also, the highly water-soluble, toxic, persistent chemical compounds like phenols are capable of significantly disturbing air, soil, and water physicochemical characteristics (State Registry of Emissions and Polluting Sources [PRTR], 2007).

Organic phenolic compounds are highly volatile in air and contribute to the formation of the tropospheric ozone, which is very harmful for humans, animals, and crops (PRTR, 2007). In water, organic phenols are highly toxic, affecting aquatic and human organisms, in which case they may lead to respiratory and circulatory problems, and cancer, in addition to their role as contaminant precursors for groundwaters through leachates generated at sanitary landfills (Institute of Hydrology, Meteorology and Environmental Studies [IDEAM, for the acronym in Spanish for Instituto de Hidrología, Meteorología y Estudios Ambientales], 2018, pp. 232–234). Lastly, in soil, organic phenolic compounds result in a decrease in productivity and changes in the use of land, landscape affectation, and loss of macro. and microfauna (Navarro, Carmona & Font, 1996, p. 49–61). For this reason, the residues of the wine-making agri-business cause significant environmental problems and represent important economic losses for the sector, besides presenting few alternatives for exploitation and appropriate method technologies for the preparation and characterization of substances with value-added potential.

According to studies, extracts obtained from organic agro-industrial residues such as cocoa (Cuéllar, Quím & Guerrero, 2012, p. 3176), tropical fruit seeds (Noguera et al., 2017, p. 33–34), and thyme (Mejía et al., 2017, p. 45) have been shown to contain substances potentially active against bacteria which were also found in wine extracts, presenting an antibacterial activity of 10% to different bacteria (Berradre et al., 2016, p. 40).

Antibacterial activity is the execution of a lethal action for bacteria (bactericides) or a transient inhibition of bacterial (bacteriostatic) growth (Calvo & Martínez, 2009, p. 45). Most compounds with antimicrobial activity have been shown to be phenolic compounds, terpenes, aliphatic alcohols, aldehydes, ketones, acids, and isoflavonoids, all natural antioxidants known to be responsible for the antibacterial, antiviral, anti-inflammatory, and anti-quorum sensing effects (Espinoza et al., 2015, p. 39; Brango, 2011, pp. 18–22), particularly the flavonoids, flavonols, catechins, anthocyanins, and polyflavonoids.

It is worth noting that many pathogenic microorganisms have a mechanism that allows communication between bacteria called quorum sensing (QS), which regulates a diverse range of physiological activities such as defense against the host immune system and against antibiotic treatment (Miller & Bassler, 2001, p. 165).

In accordance with the foregoing and in favor of looking for alternative uses, it is proposed to test the potential of substances extracted from residues of Isabella grape pomace using subcritical water (SW) regarding the antibacterial activity for stopping or slowing down the spread of strains of certain pathogens such as *Escherichia coli* and *Staphylococcus aureus*, as these microorganisms generate resistance to mechanisms (antibiotics) aiming to prevent proliferation (Bantawa et al., 2019). Likewise, analysis of the Isabella grape pomace extracts anti-QS activity obtained by pressurized liquid extraction (PLE) and supercritical fluid extraction (SFE) against the Gram-negative bacterium *Chromobacterium violaceum* was performed.

Thus, the research project focuses on presenting an alternative use for the wine-making agri-business residues (Isabella grape pomace) from extracting substances potentially active against bacteria and with anti-QS activity to provoke a positive impact on the environment and production chains, as well as adding value in the future as this represents a very applicable industrial alternative in fields including the pharmaceutical, food, and cosmetics sectors, among others.

## THEORETICAL FRAMEWORK

### QUORUM SENSING

Over 300 years ago, in the field of microbiology, the concept that bacteria were 'social' organisms with a main objective that lies in dividing to generate new bacteria was developed. However, it has been documented for more than 60 years that bacteria have a joint bacterial behavior that generates mutually interacting, complex dynamic systems and due to that interaction bacteria coexist, cooperate, compete, and exchange information in a coordinated manner (March & Eiros, 2013, p. 353). Thus, a method of bacterial communication is QS, a mechanism described in 1972 which was discovered in two species of sea bacteria, *Vibrio fischeri* and *Vibrio harveyi*, which emit luminescence only upon reaching a high-density cell population in response to the accumulation of secreted autoinductive signaling molecules (Miller & Bassler, 2001, pp. 166–167).

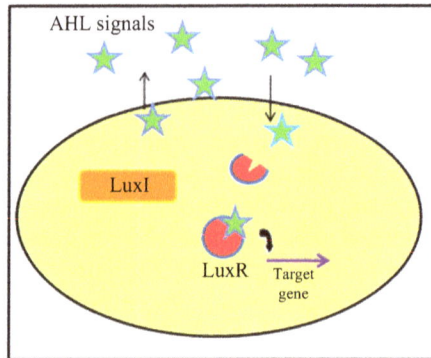

**FIGURE 6.1**   Graphic representation of the process for detecting QS in Gram-negative bacteria.

Quorum sensing detection is therefore a stimuli and response system related to the bacterial cell population density that regulates gene expression (Lek et al., 2013, p. 6218). Processes derived from this regulation include symbiosis, virulence, competition, conjugation, antibiotic production, motility, sporulation, and biofilm formation.

Quorum sensing-sensitive bacteria produce, spread, detect, and respond to chemical signaling molecules known as autoinducers that increase their concentration based on cell density. Gram-negative and Gram-positive bacteria are included among these (Lek et al., 2013, p. 6218).

For quorum detection, Gram-negative bacteria use N-acyl homoserine lactone (AHL) auto-inducing signaling molecules. The three core components of any AHL-based QS detection system are the LuxI-type synthase molecule, the AHL signaling molecule, and the LuxR-type receptor protein (Figure 6.1) (Yada et al., 2015, pp. 67–69). On the other hand, Gram-positive bacteria are found that use the oligopeptides as autoinducers.

In low-density-grown Gram-negative bacteria, AHL molecules are actively transported from the external milieu into the cytoplasm by an ATP-dependent transport process. For high concentrations, transport is carried out by passive diffusion, as the AHL concentration reaches the threshold (quorum state) the AHL autoinducing molecules interact with a regulatory protein R generally known as a transcriptional regulator. The R–AHL complex binds to the target gene(s) promoter and initiates a bacterial density-coupled transcription (Bouyahya, 2017, p. 730).

## ANTI-QS ACTIVITY

The anti-QS activity consists of blocking the QS at different stages. The QS can be inhibited in three ways: halting the production of signaling molecules so that they are not synthesized by the LuxI-encoded AHL synthase, degrading the signaling molecule, and preventing the signaling molecule from binding to its LuxR signal receptor (Yada et al., 2015, pp. 67–69). This activity may be triggered by natural products from plant compounds which typically target the bacterial QS system through the three ways mentioned above (Lek et al., 2013, p. 6222).

Disruption of QS signals can also be due to environmental factors such as increased pH, which leads to a breakdown of the autoinducing molecule lactonic ring; this mechanism has been used by plants infected by *Erwinia carotovora*. Another factor is increased temperature, which leads to the rupture of the AHL lactonic ring in *Yersinia pseudotuberculosis*, thus downregulating the expression of genes that encode flagella (March and Eiros, 2013, p. 355). Data reported from previous studies allow the identification of another approach with alternatives suitable for inhibiting QS.

Biosensors such as *E. coli* [pSB401], *E. coli* [pSB1075], and *C. violaceum* CV026 are used in studies allusive to the described topic because they induce QS features such as bioluminescence and violacein production, adapted to be able to quantify the anti-QS capacity of compounds or extracts to the relevance of inhibition (Lek et al., 2013, p. 6220).

Consequently, from the above stated facts, QS inhibition has been an attractive target for developing new measures in the pharmaceutical industry in treatments of infectious conditions since the development of resistant pathogens due to no pressure being imposed on microbe mortality is unlikely.

## CHROMOBACTERIUM VIOLACEUM

*C. violaceum* was considered a human pathogen from 1881 until 1927. Such an agent is not considered normal flora in humans or animals (Shenoy et al., 2002, p. 363) and is a soil and water saprophyte organism found in tropical and subtropical areas (Guevara et al., 2007, p. 402) as isolates of very low incidence despite having a very high mortality for humans.

*Chromobacterium violaceum* is a low-virulence agent but when it manages to establish infection it can lead to septicemia; it usually is initiated by the exposure of open skin to *C. violaceum*-contaminated soil or water, causing symptoms including fever, vomiting, blisters, lymphadenopathies, pneumonia, and visceral abscesses in the liver, spleen, lungs, and brain (Herrera et al., 2005, p. 5).

The shape of the Gram-negative bacterium *C. violaceum* is that of a rod. It is a facultative anaerobic (grows in the presence or absence of oxygen), fermentative (produces hydrogen and carbon dioxide), and oxidase-positive microorganism characterized by being mobile, presenting a polar flagella, and a growth temperature range between 15 and 40°C. *C. violaceum* produces a non-diffusible violet pigment (Herrera et al., 2005, p. 5) called violacein, which is soluble in ethanol and non-soluble in water and chloroform.

Currently, some of the virulence mechanisms, such as the production of molecules including, but not limited to, hemolysins, cytolysins, elastases, and violacein are known. Violacein has become og increasing interest within the scientific community and has been suggested to play a major role in *C. violaceum* QS since regulation of its synthesis is induced by N-hexaonyl homoserine lactone (HHL), a quorum sensing signaling molecule whose concentration in the medium stimulates an increase of the bacteria population (Guevara et al., 2007, p. 404).

### Antimicrobial Activity

The antimicrobial activity is carried out by compounds that inhibit the growth or cause the death of bacteria and other microorganisms that occur in large diversity and, in turn, depend on the targets these compounds are directed to (Calvo & Martínez, 2009, p. 44). Species of microorganisms such as bacteria, fungi, and actinomycetes produce chemicals that suppress the growth of other microorganisms and that, combined with other substances, can produce synthetic compounds called antibiotics (Brugueras & García, 1998, pp. 348–349).

The antimicrobial activity mechanisms of action can occur in five different ways: by inhibiting bacterial wall synthesis, altering the cytoplasmic membrane, inhibiting protein synthesis, altering nucleic acid metabolism or structure, and blocking the synthesis of metabolic factors (Calvo & Martínez, 2009, p. 46). On the other hand, bacteria are unicellular prokaryotic microorganisms that differ in their ability to retain a dye and are classified as Gram-positive or Gram-negative (Brugueras & García, 1998, pp. 348–349).

Two bacteria are most often used for antibacterial activity studies: *Escherichia coli* discovered in 1885 by Dr. Theodor Escherich (Huerta, 2020, p. 1) was described as a Gram-negative, non-sporulating, facultative anaerobic rod of the *Escherichia* genus Enterobacteriaceae family (Rodríguez, 2002, p. 465) and *Staphylococcus*, described for the first time in 1880 by the physician Alexander Ogston (Phaissa, 2009, p. 15) as an immobile, facultative, Gram-positive, cocci-shaped, non-spore-forming anaerobic bacteria found as normal microbiota in living beings that typically have a yellowish color due to the production of a carotenoid pigment (Phaissa, 2009, p. 11).

The bacteria mentioned above are a strong pillar of bacteria and microbiology research as they are easy to cultivate, manipulate, and characterize, and also are useful for the purpose of learning about bacterial control mechanisms (Phaissa, 2009, p. 11–18; Huertas, 2020).

## METHODOLOGY

### Sample Collection and Preparation

The samples that were used in the project were provided by the Valle del Cauca wine-making agri-business Casa Grajales (Grajales Industry Group) and are constituted by the Isabella grape (*Vitis labrusca* L) bagasse or pomace generated during the production of fermented beverages. The sample was subjected to manual cleaning to remove any material other than peels and seeds. Grinding of the sample was subsequently carried out and it was subjected to sieving and granulometry until a size of between 0.2 and 0.5 mm was reached.

### Extract Collection

To obtain extracts, three types of extraction were considered, including:

- Supercritical fluid extraction (SFE): For SFE, pure supercritical $CO_2$ (SC $CO_2$) and ethanol-spiked SC $CO_2$ as co-solvent are used at different pressures and temperatures. The methodology proposed by Castro et al. (Castro et al., 2010; Castro et al., 2011a, 2011b) was followed for this purpose.

- Pressurized liquid extraction (PLE): The extracts were obtained by means of PLE using as solvent mixtures of water:ethanol modifying the temperature, composition of the mixture and pH following the methodology proposed by Ballesteros (Ballesteros, 2015), Mustafa and Turner (Mustafa & Turner, 2011, pp. 8–18), and Christen and Kaufmann (Christen & Kaufmann, 2014, pp. 15–16).
- Subcritical water extraction (SWE): Extracts were obtained by means of subcritical water following the methodology described by Yang (Yang et al., 2013, p. 121) and Tian (Tian et al., 2017, p. 3), where a temperature variation above water boiling point occurred at five different levels and constant pressure above 100 bar.

## ANTI-QS ACTIVITY TESTS

Once the extracts were obtained by PLE (Table 6.1), SFE (Table 6.2), and SWE (Table 6.3), a standard solution was prepared by dissolving about 100 mg of dry extract into a water:ethanol (50% each) solution until a concentration of 10 mg/mL was reached. For this purpose, the methodology proposed by Hatamnia, Abbaspour, and Darvishzadeh (2014, pp. 307–308) and Ng et al. (2019, p. 124) was followed.

## EVALUATION OF THE ANTI-QUORUM SENSING ACTIVITY

Once the extracts had been diluted, their anti-QS activity was determined following the methodology proposed by Viola et al. (Viola et al., 2020, pp. 153–154), where the agar wells technique was used to measure the inhibitory activity of Isabella grape (*Vitis labrusca*, L) pomace extract. The bacterium *C. violaceum* ATCC 31532 was supplied by the National University of Colombia research group of Bacterial

**TABLE 6.1**
**Conditions for pressurized liquid extraction (PLE)**

| PLE extract | %EtOH | %H$_2$O | Temperature (°C) | Mass (mg) |
|---|---|---|---|---|
| 1 | 40 | 60 | 60 | 97.4 |
| 2 | 60 | 40 | 60 | 96.9 |
| 3 | 40 | 60 | 100 | 98.8 |
| 4 | 60 | 40 | 100 | 100.1 |
| 5 | 50 | 50 | 51.7 | 94.1 |
| 6 | 50 | 50 | 108.3 | 100.9 |
| 7 | 35.9 | 64.1 | 80 | 99.7 |
| 8 | 64.1 | 35.9 | 80 | 105.0 |
| 9 | 50 | 50 | 80 | 93.4 |
| 10 | 50 | 50 | 80 | 109.9 |
| 11 | 50 | 50 | 80 | 93.0 |
| 12 | 50 | 50 | 80 | 102.1 |

**TABLE 6.2**
**Conditions for supercritical fluid extraction (SFE)**

| SFE extract | %Co. EtOH | Concentration (v/v) | Mass (mg) |
|---|---|---|---|
| 13 | 5 | 57 | 98.025 |
| 14 | 10 | 57 | 96.93 |
| 15 | 15 | 57 | 107.195 |
| Pressure | 32.1 MPa | | |
| Temperature | 58.4 °C | | |

**TABLE 6.3**
**Conditions for subcritical water extraction**

| SWE extract | Temperature (°C) |
|---|---|
| P1 | 100 |
| P2 | 125 |
| P3 | 150 |
| P4 | 175 |
| P5 | 200 |
| Pressure | 100 bar |

Communities and Communication to conduct the analyses previously mentioned. *Chromobacterium violaceum* was cultured in Luria Bertani and soybean trypticase culture media. For inhibition tests, bacteria were incubated in trypticase soy broth for 24 h at 28°C in an orbital shaker at 150 rpm. Subsequently, 110 μL of grown *C. violaceum* were added to a LB agar-containing Petri dish and using a glass loop *C. violaceum* was plated, and wells of approximately 1 cm in diameter were punched using a puncher. Concentrations of 0.5 mg (50 μL), 0.75 mg (75 μL), and 1 mg (100 μL) of each extract were added into the wells. Assays were set up in triplicate in the same way for each concentration. A blank was included in the test with the same amounts described above and prepared in a 50% water and 50% ethanol solution. Figure 6.2 shows the distribution of extracts and blanks in the Petri dish at the set-up of the anti-QS activity testing.

Plates were left at room temperature and exposed to sunlight for 3 days once the wells were seeded with extracts to favor the expression of violacein. For ascertainment of extract activity, controls were performed at 18 h, 24 h, 32 h and 48 h. The purpose of these incubation times was to obtain a halo of inhibition around the well which is cloudy and where no violacein synthesis is displayed, unlike the no extract or blank spots on the plate, so as to determine what extracts exhibit QS inhibition to *C. violaceum*. A digital caliper was the tool used to measure the inhibition halos.

| DISH #1 | DISH #2 | DISH #3 |
|---|---|---|
| C: 75µL<br>C: 50µL<br>B: 50µL | C: 50µL<br>C: 75µL<br>B: 75µL | C: 100µL<br>C: 75µL<br>B: 75µL |

| DISH #4 | DISH #5 | DISH #6 |
|---|---|---|
| C: 100µL<br>C: 75µL<br>B: 100µL | C: 50µL<br>C: 100µL<br>B: 50µL | C: 100µL<br>C: 50µL<br>B: 100µL |

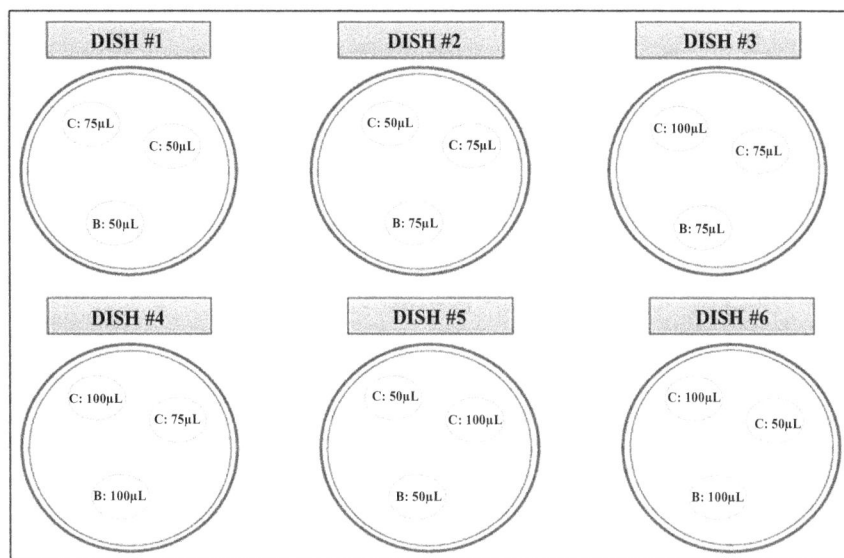

**FIGURE 6.2**   Distribution of concentrations at the set-up of anti-QS activity testing.

*Note:* Assay performed by the agar wells method.

## ANTIBACTERIAL ACTIVITY TESTS

The extracts used were those obtained by subcritical water extraction from which a standard solution was prepared by dissolving each extract in ethanol until a concentration of 40 mg/mL or 50 mg/mL was reached.

## EVALUATION OF ANTIBACTERIAL ACTIVITY

Once the extracts were diluted, antibacterial activity following the methodology proposed by Balouri, Sidiki, and Koraichi (Balouri, Sadiki, and Koraichi, 2016, pp. 72–73) and the Spanish Society of Infectious Diseases and Clinical Microbiology (Picazo, 2000, pp. 4–5) was determined by means of the agar disc diffusion technique to measure the antibacterial activity of the Isabella grape (*Vitis labrusca* L) pomace extract.

*Escherichia coli* (ATCC 25922) and *S. aureus* (ATCC 29213) bacteria to conduct the testing were procured from Microbiologics. These bacteria were cultured in Luria Bertani and soybean trypticase culture media. The methodology proposed by Ng et al. (2019, pp. 124–125) was used as a reference for antibacterial activity testing. Thus, bacteria were incubated in trypticase soy broth separately for 24 h at 28°C in an orbital shaker at 150 rpm. Subsequently, 100 µL of *E. coli* grown in liquid medium were plated onto a Petri dish containing LB agar and tested by means of the disc diffusion method. The procedure for *S. aureus* was carried out following the same

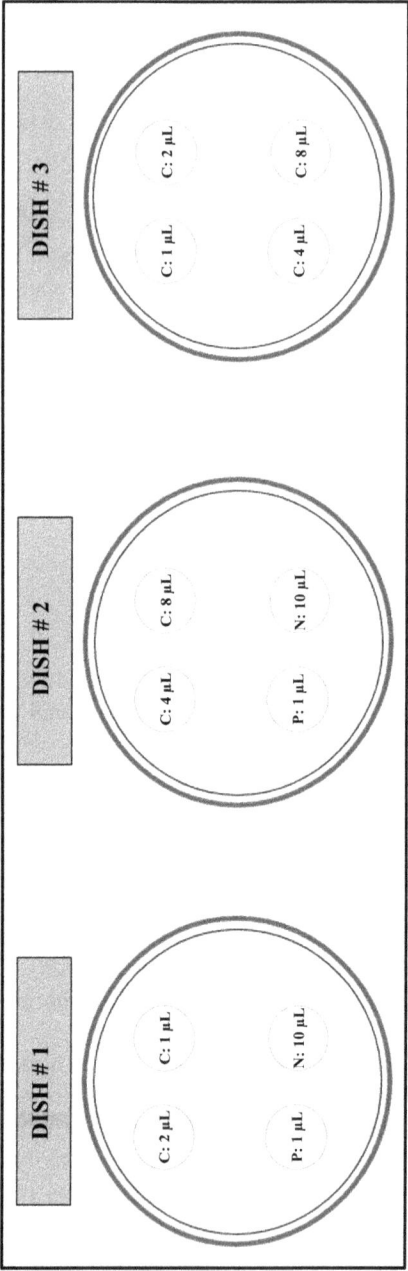

**FIGURE 6.3** Distribution of concentrations for the antibacterial activity testing set-up on extracts using a 50 mg/mL standard solution.

*Note:* The distribution of concentrations for the antibacterial activity testing set-up by means of the disk diffusion method on extracts whose standard solution is 50 mg/mL is shown in the figure.

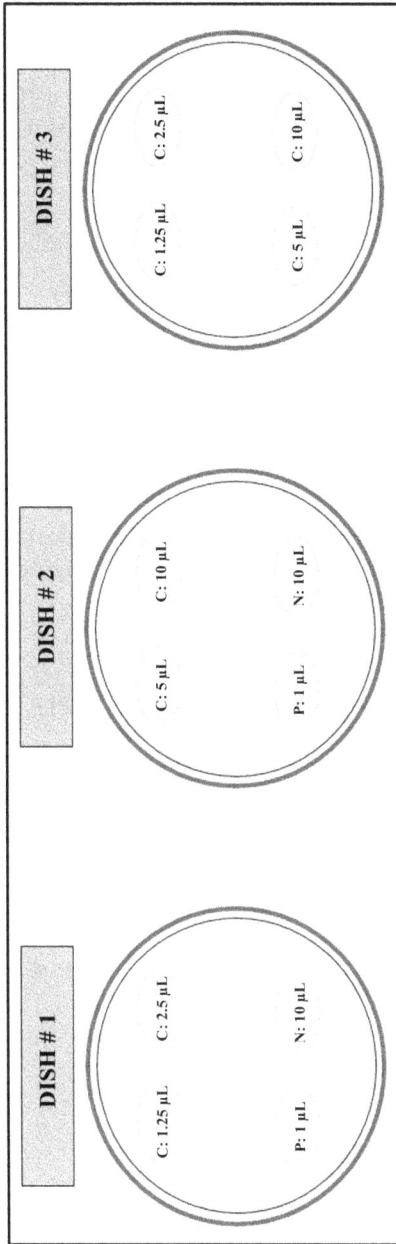

**FIGURE 6.4** Distribution of concentrations for the antibacterial activity testing set-up on extracts using a 40 mg/mL standard solution.

*Note:* The distribution of concentrations for the antibacterial activity testing set-up by means of the disk diffusion method on extracts whose standard solution is 40 mg/mL is shown in the figure.

protocol. Concentrations of 0.05 mg (1 and 1.25 µL), 0.1 mg (2 and 2.5 µL), 0.2 mg (4 and 5 µL), and 0.4 mg (8 and 10 µL) were added to the disks with the first volume corresponding to those extracts whose standard solution was at a concentration of 50 mg/mL and the second at a concentration of 40 mg/mL. The assays were set up in duplicate both for the two concentrations as well as for controls. A positive control was considered in the experiments which corresponded to ampicillin at a concentration of 10 µg/disk and a negative control using a 96% ethanol 10 µL aliquot. The distribution considered on the Petri dishes for the concentrations as the antibacterial activity testing was set up as illustrated in Figures 6.3 and 6.4.

## RESULTS

### EXTRACTS OBTAINED BY PRESSURIZED LIQUID EXTRACTION (PLE)

Extracts obtained using PLE resulted in anti-QS activity effects against the bacterium *C. violaceum*. The results are shown in Table 6.4, where mean diameters for inhibition in millimeters at areas around the wells are presented.

Graphical evidence of results shown in Table 6.4 is presented in Figures 6.5 and 6.6.

According to Table 6.4 and Figure 6.5, extract 2 anti-QS activity occurs only at a concentration of 1.00 mg/well. According to Table 6.4 and Figure 6.6, extract 5 anti-QS activity was observed at the tested concentrations of 0.50 mg/well, 0.75 mg/well,

### TABLE 6.4
### Anti-quorum sensing activity of extracts obtained by PLE

| PLE extract | Negative control halo (mm) | Halo obtained at used concentrations (mm) | | |
|---|---|---|---|---|
| | | 0.5 (mg/well) | 0.75 (mg/well) | 1.0 (mg/well) |
| 1 | – | – | – | – |
| 2 | – | – | – | 16.31± 0.85 |
| 3 | – | – | – | – |
| 4 | – | – | – | – |
| 5 | – | 14.56 ± 0.00 | 15.22 ± 2.35 | 16.69 ± 3.44 |
| 6 | – | – | – | – |
| 7 | – | – | – | – |
| 8 | – | – | – | – |
| 9 | – | – | – | – |
| 10 | – | – | – | – |
| 11 | – | – | – | – |
| 12 | – | – | – | – |

*Note:* The anti-QS activity of Isabella grape (*Vitis labrusca* L.) extracts obtained by means of the PLE technology against the bacterium *C. violaceum* are represented in Table 6.4 and are estimated from the mean diameters of the areas that present communication inhibition.

and 1.00 mg/well. On the other hand, no anti-QS activity was elicited from extracts 1, 4, 6, 7, 8, 9, 10, 11, and 12.

## EXTRACTS OBTAINED BY SUPERCRITICAL FLUID (SFE)

Results from well diffusion testing for extracts obtained using SFE are shown in Table 6.5, where the use of concentrations of 0.50, 0.75, and 1.00 mg/well is shown demonstrating no anti-QS activity against the bacterium *C. violaceum.*

## EXTRACTS OBTAINED BY SUBCRITICAL WATER (SWE)

Extracts obtained using subcritical water by PLE resulted in effects of antibacterial activity against the bacterium *E. coli.* However, against the bacterium *S. aureus*, no antibacterial activity was reported for the extracts. The results are shown in Table 6.6, where the mean diameters of growth inhibition in areas around the disks in millimeters are presented.

Graphical evidence of results shown in Table 6.6 is presented in Figure 6.7 and 6.8.

As a result, antibacterial activity in extracts P2 and P5 is obtained, which were collected at temperatures of 125°C and 200°C, respectively, using subcritical water. According to Table 6.6 and Figure 6.7, antibacterial activity at the concentrations of 0.05 mg/disk (7.10 mm), 0.1 mg/disk (9.12 mm), 0.2 mg/disk (8.87 mm), and 0.4 mg/disk (8.83 mm) could be found for extract P2. According to Table 6.6 and

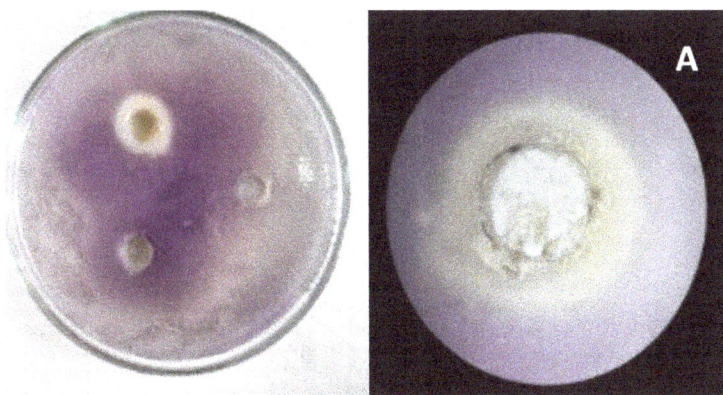

**FIGURE 6.5**   Anti-quorum sensing activity of PLE obtained extract 2. Photograph by the authors.

*Note:* Concentration of 1 mg/well exhibiting an inhibition halo of 16.31 mm. Images captured through a stereoscope Stemi DV4 model (ZEISS).

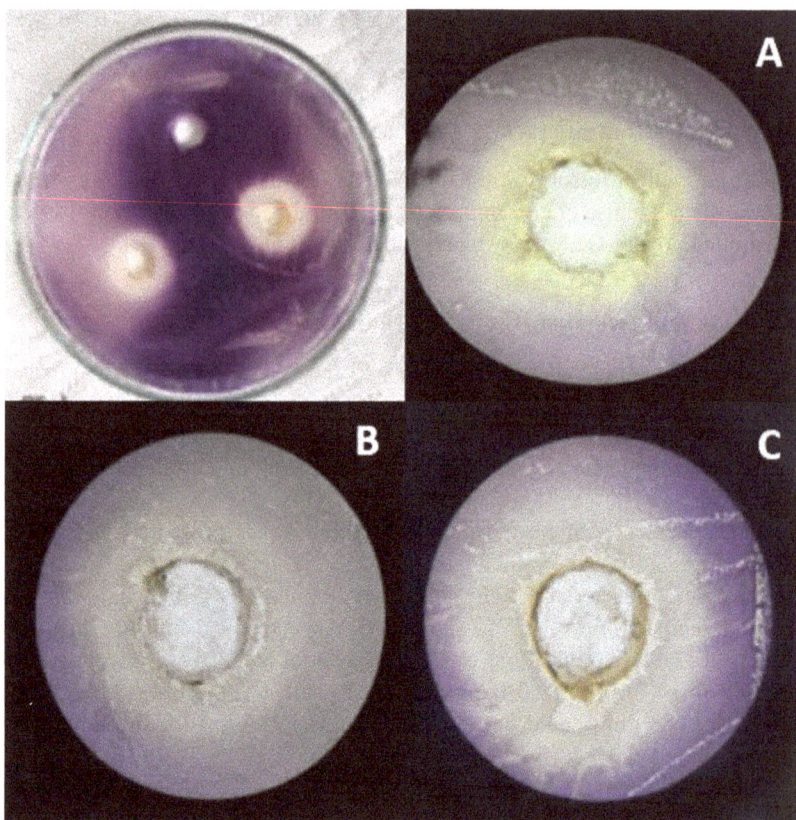

**FIGURE 6.6** Anti-quorum sensing activity of PLE obtained extract 5. Photograph by the authors.

*Note:* (A) Concentration of 0.5 mg/well exhibiting an inhibition halo of 14.56 mm. (B) Concentration of 0.75 mg/well exhibiting an inhibition halo of 15.22 mm. (C) concentration of 1 mg/well exhibiting an inhibition halo of 16.69 mm. Images captured using a stereoscope Stemi DV4 model (ZEISS).

**TABLE 6.5**
**Anti-quorum sensing activity by SFE obtained extracts**

| SFE extract | Negative control halo (mm) | Halo obtained at used concentrations (mm) | | |
| --- | --- | --- | --- | --- |
| | | 0.5 (mg/well) | 0.75 (mg/well) | 1.0 (mg/well) |
| 13 | – | – | – | – |
| 14 | – | – | – | – |
| 15 | – | – | – | – |

**TABLE 6.6**
**Antibacterial activity of SWE obtained extracts**

| SWE extract | Bacterial strain | Positive control halo 10 mg/well (mm) | Negative control halo (mm) | Halo obtained at used concentrations (mm) | | | |
|---|---|---|---|---|---|---|---|
| | | | | 0.05 (mg/well) | 0.10 (mg/well) | 0.20 (mg/well) | 0.40 (mg/well) |
| P1 | S. aureus | 11.83 ± 0.00 | – | – | – | – | – |
| | E. coli | 14.70 ± 0.00 | – | – | – | – | – |
| P2 | S. aureus | 11.82 ± 1.52 | – | – | – | – | – |
| | E. coli | 13.55 ± 1.25 | – | 7.10 ± 0.00 | 9.12 ± 2.28 | 8.87 ± 0.00 | 8.83 ± 1.92 |
| P3 | S. aureus | 16.49 ± 0.00 | – | – | – | – | – |
| | E. coli | 15.36 ± 0.23 | – | – | – | – | – |
| P4 | S. aureus | – | – | – | – | – | – |
| | E. coli | 15.00 ± 0.00 | – | – | – | – | – |
| P5 | S. aureus | 16.35 ± 2.82 | – | – | – | – | – |
| | E. coli | 12.88 ± 2.38 | – | 8.29 ± 0.27 | 8.13 ± 0.01 | 8.01 ± 1.05 | 8.20 ± 0.78 |

*Note:* Mean diameters of areas showing growth inhibition are presented in Table 6.6. Negative control: 96% ethanol from which a 10 µL/disk aliquot was taken. Positive control: ampicillin at a concentration of 10 µg/disk.

Figure 6.8, extract P5 antibacterial activity at the tested concentrations of 0.05 mg/disk (8.29 mm), 0.1 mg/disk (8.13 mm), 0.2 mg/disk (8.01 mm), and 0.4 mg/disk (8.20 mm) was evidenced. On the other hand, no antibacterial activity was seen for extracts P1, P3, and P4.

## DISCUSSION

For assessment of anti-QS activity, 15 extracts of Isabella grape (*Vitis labrusca*, L.) pomace were tested against the bacterium *C. violaceum* (ATCC 31532) from which 12 extracts were obtained using PLE and three using SFE at different temperature and pressure conditions. The anti-QS activity of the extracts may be directly linked to the total content of phenolic compounds present in the samples analyzed according to literature queries (Borges et al., 2014, p. 183; Choo, Rukayadi & Hwang, 2005, p. 640). Among these compounds the flavonoids (anthocyanins family), which are responsible for the red, blue, or violet tones of fruit skins (Gimeno, 2004, p. 81), exist. Of the extracts obtained, the highest concentrations of phenols collected using PLE were found in extracts 3, 4, and 6, described in Chapter 5. These extracts were obtained at the highest temperatures ranging from 100°C to 108.3°C. However, according to previous studies, impaired stability of extract flavonoids is shown by increasing the extraction temperature generating a higher rate of degradation (Arrazola, Herazo, and Alvis, 2014, pp. 43–44). In addition, flavonoids show QS inhibition with no disturbance of microorganism growth; higher quantities of flavonoids may be found in polar extracts (Brango, 2011,

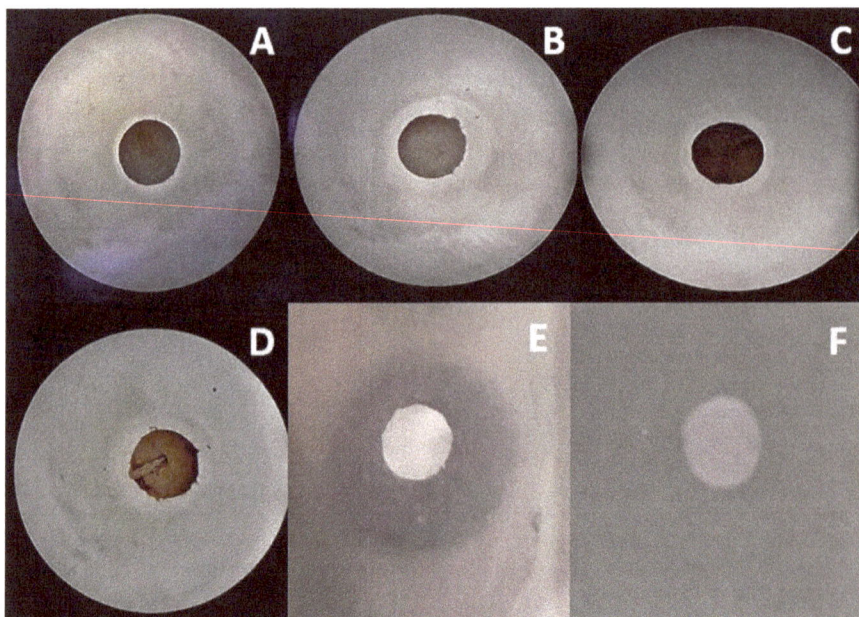

**FIGURE 6.7** Antibacterial activity of subcritical water obtained extract P2. Photograph by the authors.

*Note:* At the center of the image the sensidisk and the inhibition halo obtained around it can be found. (A0 Concentration of 0.05 mg/disk exhibiting an inhibition halo of 7.10 mm. (B) Concentration of 0.10 mg/disk exhibiting an inhibition halo of 9.12 mm. (C) Concentration of 0.20 mg/disk exhibiting an inhibition halo of 8.87 mm. (D) Concentration of 0.40 mg/disk exhibiting an inhibition halo of 8.83 mm. (E) Positive control, ampicillin at a concentration of 10 µg/disk presenting an inhibition halo of 13.55 mm. (F) Negative control consisting of 96% ethanol in a 10 µL/disk aliquot presenting an inhibition halo of 0.00 mm. Images captured using a stereoscope Stemi DV4 model (ZEISS).

p. 43). Therefore, it can be concluded that at higher extraction temperatures, the lower (or even absence thereto) is the inhibitory activity displayed by extracts. The findings above are evidenced by results from this study as PLE-collected extracts 2 and 5 were obtained at a lower extraction temperature (51.7°C and 60°C) using a flavonoid concentration of 590.36 µg quercetin/g extract and 651.76 µg quercetin/g extract, respectively.

It should be noted that, according to previous studies conducted by Viola et al. (Viola et al., 2020, pp. 163–164), extracts from wine-making waste have been shown to have anti-QS activity in concentrations of 2.50 mg and 5.00 mg against the bacterium *C. violaceum* (ATCC 12472 and CV 026). Extracts were collected by an ethyl acetate solid–liquid method and used organic solvents. On the other hand, studies carried out by Sheng et al. (Sheng et al., 2016, pp. 5–6) showed that grape seed extracts acquired from Optipure (Los Angeles, CA) contained 95% of total flavonoids and were prepared in a 40 mg/mL, 10% ethanol solution for the extraction of the

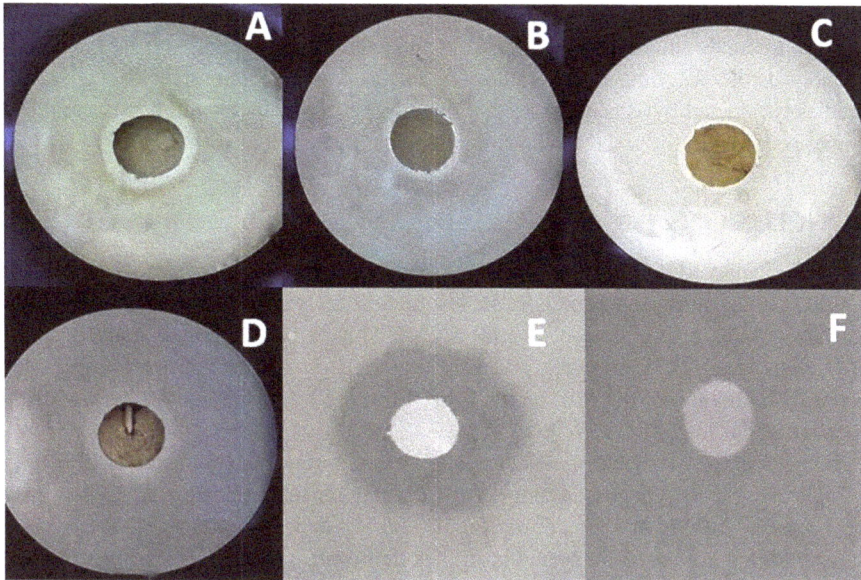

**FIGURE 6.8** Antibacterial activity of subcritical water obtained extract P5. Photograph by the authors.

*Note:* At the center of the image the sensidisk and the inhibition halo obtained around it can be found. (A) Concentration of 0.05 mg/disk exhibiting an inhibition halo of 8.29 mm. (B) Concentration of 0.10 mg/disk exhibiting an inhibition halo of 8.13 mm. (C) Concentration of 0.20 mg/disk exhibiting an inhibition halo of 8.01 mm. (D) Concentration of 0.40 mg/disk exhibiting an inhibition halo of 8.20 mm. (E) Positive control, ampicillin at a concentration of 10 µg/disk presenting an inhibition halo of 12.88 mm. (F) Negative control consisting of 96% ethanol in a 10 µL/disk aliquot presenting an inhibition halo of 0.00 mm. Images captured using a stereoscope Stemi DV4 model (ZEISS).

total phenolic compounds, thus demonstrating that the grape seed oil exhibits anti-QS activity against a bacteria cocktail called *E. coli* "top-six" non-O157 STEC at concentrations of 0.50 mg/mL for the *E. coli* strain O26:H11, and 1–4 mg/mL for the non-O157 STEC strain. However, no studies on extracts obtained by PLE and SFE were found using the Isabella grape (*Vitis labrusca* L.) variety where its anti-QS potential against the bacterium *C. violaceum* ATCC 31532 was tested. The results obtained in this study provide the basis for analyzing the applicability of extracts as substitutes in the use of antifouling paints responsible for generating a toxic and devastating effect on marine flora and fauna (Barreiro, Quintela and Ruiz, 2004, pp. 14; Rodriguez et al., 2009, p. 2).

For the assessment of antibacterial activity, five extracts from Isabella grape (*Vitis labrusca*, L.) pomace were tested against the bacteria *E. coli* (ATCC 25922) and *S. aureus* (ATCC 29213), which were obtained by SWE at different temperature conditions and constant pressure, as recorded in detail in Table 6.3. Different types of grape extracts have demonstrated potential antibacterial activity as evidenced by different results obtained from previous studies. In the context of these assessments, the main compounds and molecules whose potential activity is predominant are flavonoids, phenolic acids, catechins and proanthocyanidins, and anthocyanins (Silván et al., 2013). Studies conducted by Hashim state that the effect of proanthocyanidin extracted from grape seeds is antimicrobial on the strains tested (Hashim et al., 2020, p. 200).

## CONCLUSIONS

Extracts displayed an anti-QS activity in the well diffusion bioassay where a clear, white, opaque inhibition zone is observed in cultures where the *C. violaceum* strain was plated at a concentration of 1.00 mg/well for extract 2, and at concentrations of 0.50 mg/well, 0.75 mg/well, and 1.00 mg/well for extract 5. Moreover, extracts obtained by means of the SFE technology showed no anti-QS activity at any of the concentrations tested.

Isabella grape (*Vitis labrusca* L.) pomace extract obtained by means of PLE technology using 100% degassed and distilled water on the bacterium *E. coli* showed antibacterial activity at a concentration of 0.05 mg/disk for extracts P2 and P5. The latter findings are based on previous studies demonstrating that, regarding chemical composition, grape seed oil is rich in phenolic compounds to which antibacterial properties have been attributed (Burt, 2004, p. 225). On the other hand, no antibacterial activity resulting from exposure to the bacterium *S. aureus* was reported for the remaining extracts collected by means of the PLE technology using subcritical water.

An alternative use for agro-industrial waste as a promising source for extraction and isolation of bioactive compounds capable of intervening in bacterial virulence processes is presented by this research. Likewise, the applicability for the extracts obtained in the pharmaceutical and cosmetic industries, or as additives in the food industry is demonstrated.

## GLOSSARY

**Antibacterial:** An antibacterial is a substance or compound that kills or reduces the spread of bacteria only.

**Antimicrobial:** An antimicrobial is a substance or compound that kills or reduces the spread of a broad spectrum of microorganisms: this includes bacteria, mold, fungi, and even viruses.

**Chromobacterium violaceum:** This is a Gram-negative bacillus that is facultative anaerobic, motile, and oxidase-positive (Yang & Li, 2011, p. 435). It is commonly found in tropical aquatic environments and natural soils and is sensitive to temperature.

**Escherichia coli:** This is a Gram-negative, non-sporulating, facultative anaerobic bacillus of the Enterobacteriaceae family, *Escherichia* tribe (Rodríguez, 2002, p. 465). It has been recognized as a human pathogen since its discovery and it inhabits the intestinal tracts of humans and warm-blooded animals (Barreto, 1997, p. 2)

**Fruit sector:** According to the dictionary of the Royal Spanish Academy, the fruit sector refers to the set of techniques and knowledge related to the cultivation of fruit trees.

**Pomace:** The word pomace comes from the common Latin *voluculum*, which means sheath or little skin that surrounds something, which by reducing syllables reached the form pomace, thus converting it to the definition that is currently known as pomace referring to the pomace or skin of the grape, after being crushed to obtain the must (Escobar, 2018).

**Quorum sensing:** This was discovered more than 25 years ago in two species of marine bacteria, *Vibrio fischeri* and *Vibrio harveyi*, and it refers to the regulation of gene expression in response to cell–cell signaling (Miller & Bassler, 2001, 166–167).

**Staphylococcus aureus:** This is an immobile, Gram-positive, coccus-shaped, facultative anaerobic bacterium that does not form spores and is found in the normal microbiota of living beings (Bhatia & Zahoor, 2007, p. 188).

**Wine-growing:** According to the dictionary of the Royal Spanish Academy, the wine industry refers to everything related to the manufacture of wine or a person who has a vineyard estate and is practical in its cultivation.

**Vitis labrusca L.:** This is a variety of Isabella tinta grape that, over the years, has optimally adapted to the soil and climate conditions of the Colombian region, especially in Norte de Santander and Boyacá (Hernández, Trujillo & Durán, 2011, p. 17–18).

## ACKNOWLEDGMENTS

To the Universidad Libre for funding the research project entitled "Exploitation and valorization of waste generated by the Colombian wine-making agri-business through obtaining additives potentially applicable in food, cosmetic, and pharmaceutical industry using green technologies," a scientific endeavor of which this research is part. To the Department of Biology's Bacterial Communities and Communication research group of the National University of Colombia at Bogotá for providing the biosensor *C. violaceum* (ATCC 31532).

## REFERENCES

Arrazola, G. Herazo, I. & Alvis, A. (2014). Obtención y Evaluación de la Estabilidad de Antocianinas de Berenjena (*Solanum melongena* L.) en Bebidas. Scielo, 25 (3), 43–49. http://dx.doi.org/10.4067/S0718-07642014000300007

Ballesteros, D. (2015). Estudio comparativo sobre la obtención de extractos con actividad citotóxica a partir de residuos frutícolas. Trabajo de grado. Universidad Nacional de Colombia. Facultad de Ciencias. Departamento de Química. Bogotá. www.semantic scholar.org/paper/Estudio-comparativo-sobre-la-obtenci%C3%B3n-de-extractos-Vivas/0924a62f240aa415381111f3041b9c434a5a625c

Balouri, M.; Sadiki, M.; Koraichi, S. (2016). Methods for in vitro evaluating antimicrobial activity: A review. Journal of Pharmaceutical Analysis, 6 (2), 72–77. https://doi.org/10.1016/j.jpha.2015.11.005

Bantawa, K., Sah, S. N., Subba Limbu, D., Subba, P., & Ghimire, A. (2019). Antibiotic resist-ance patterns of Staphylococcus aureus, Escherichia coli, Salmonella, Shigella and Vibrio isolated from chicken, pork, buffalo and goat meat in eastern Nepal. BMC Research Notes, 12(1), 766. https://doi.org/10.1186/s13104-019-4798-7

Barreiro, R., Quintela, M., Ruiz, J., (2004). TBT e imposex en Galicia: Los efectos de un dis-ruptor endocrino en poblaciones de gasterópodos marinos. Ecosistemas, 13(3), 13–29. www.revistaecosistemas.net/index.php/ecosistemas/article/view/196

Barreto,G.(1997). Escherichia coli: un reto después de 111 años de estudio. Revista Archivo Médico de Camagüey, 1(2), 2–3. http://scielo.sld.cu/scielo.php?script=sci_arttext&pid=S1025-02551997000200010&lng=es&nrm=iso

Beleño, I. (27 de marzo de 2018). En el sector agrícola se pierden 6 millones de toneladas de alimentos al año. www.agronegocios.co/agricultura/en-el-sector-agricola-se-pierden-6-millones-de-toneladas-de-alimentos-al-ano-2706145

Berradre, M., Vides, A., Ojeda, G., Soto, L., Sulbarán, B., Fernández, V., & Peña, J. (2016). Caracterización fisicoquímica y actividad antibacteriana del aceite de semillas de uva (Vitis vinifera) variedad Malvasía. Trabajo de grado. Universidad del Zulia. Facultad experimental de ciencias. Departamento de Química. Maracaibo, (33), 40.

Bhatia, A. & Zahoor, S. (2007). Staphylococcus Aureus Enterotoxins: A Review. Journal of Clinical and Diagnostic Research, 1(2), 188–197. http://citeseerx.ist.psu.edu/viewdoc/download?doi=10.1.1.621.3163&rep=rep1&type=pdf

Borges, A., Serra, S., Cristina Abreu, A., Saavedra, M. J., Salgado, A., & Simões, M. (2013). Evaluation of the effects of selected phytochemicals on quorum sensing inhibition and in vitro cytotoxicity. Biofouling, 30 (2), 183–195. https://doi.org/10.1080/08927 014.2013.852542

Bouyahya, A., Dakka, N., Et-Touys, A., Abrini, J., & Bakri, Y. (2017). Medicinal plant products targeting quorum sensing for combating bacterial infections. Asian Pacific Journal of Tropical Medicine, 10(8), 729–743. doi:10.1016/j.apjtm.2017.07.021

Brango, J. (2011). Búsqueda de Compuestos Inhibidores de Quorum Sensing (IQS) a Partir de Extractos de Origen Natural. Primera Fase. (Trabajo de grado). Bogotá Colombia. 20. https://repositorio.unal.edu.co/bitstream/handle/unal/10818/197496.2011.pdf?seque nce=1&isAllowed=y

Brugueras, M. García, M. (1998). Antibacterianos de acción sistémica. Parte I. Antibióticos betalactámicos. Revista Cubana de Medicina General Integral, 14(4). 348–359.http://sci elo.sld.cu/scielo.php?script=sci_arttext&pid=S0864-21251998000400008

Burt, S. (2004). Essential oils: their antibacterial properties and potential applications in foods – a review. International Journal of Food Microbiology, 94, 223–253. https://pub med.ncbi.nlm.nih.gov/15246235/

Calvo, J. Martinez, L. (2009). Mecanismos de acción de los antimicrobianos. Enfermedades Infecciosas Microbiología Clínica, 27(1), 44–51. www.elsevier.es/es-revista-enfermeda des-infecciosas-microbiologia-clinica-28-articulo-mecanismos-accion-los-antimicrobia nos-S0213005X08000177

Castro, H., Rodríguez, L,,, Ferreira, S., Parada, F. (2010). Extraction of phenolic fraction from guava seeds (Psidium guajava L.) using supercritical carbon dioxide and co-solvents. The Journal of Supercritical Fluids, 51(3), 319–324.

Castro, H., Rodríguez, L., Parada, F. (2011a). Aprovechamiento integral de la guayaba (Psidium guajava L): IV Fracciones antagónicas de semillas de guayaba obtenidas mediante extracción con $CO_2$ supercrítico y cosolvente https://doi.org/10.1016/j.sup flu.2009.10.012

Castro, H., Rodríguez, L., Ferreira, S., Parada, F. (2011b). Guava (Psidium guajava L.) seed oil obtained with a homemade supercritical fluid extraction system using supercritical

CO2 and co-solvent. The Journal of Supercritical Fluids, 56(3), 238–242. https://doi.org/10.1016/j.supflu.2010.10.040

Chávez, Á. & Rodríguez, A. (2016). Aprovechamiento de residuos orgánicos agrícolas y forestales en Iberoamérica. Academia y Virtualidad 9 (2), 90–107. http://dx.doi.org/10.18359/ravi.2004

Choo, J., Ruykayadi, Y., Hwang, J. (2005). Inhibition of bacterial quorum sensing by vanilla extract. Letters in Applied Microbiology, 42, 637–641. doi:10.1111/j.1472-765x.2006.01928.x

Christen, P., & Kaufmann, B. (2014). New trends in the extraction of natural products: microwave-assisted extraction and pressurized liquid extraction. Encyclopedia of Analytical Chemistry, 1–27. https://doi.org/10.1002/9780470027318.a9904

Cuéllar, O., Quím, T., Guerrero, G. (2012) Actividad antibacteriana de la cáscara de cacao, *Theobroma cacao* L. Revista MVZ Córdoba, 17 (3), 3176–3186. https://doi.org/10.21897/rmvz.218

Cury, K., Aguas, Y., Martínez, A., Olivero, R. & Chams, L. (2017). Residuos agroindustriales su impacto, manejo y aprovechamiento. Revista Colombiana de Ciencia Animal, 9, 122–132. https://doi.org/10.24188/recia.v9.nS.2017.530

Departamento Administrativo Nacional de Estadística. (29 de Septiembre de 2020). Encuesta nacional agropecuaria (ENA). www.dane.gov.co/index.php/estadisticas-por-tema/agropecuario/encuesta-nacional-agropecuaria-ena#:~:text=La%20Encuesta%20Nacional%20Agropecuaria%20%E2%80%93%20ENA,%25

Escobar, S. (2018). Residuos de almazara. Segunda Parte: El orujo de oliva. www.antojodelsur.com/residuos-almazar-el-orujo-de-oliva/#:~:text=Originalmente%20la%20palabra%20orujo%20solo,aplastada%20para%20obtener%20el%20mosto.&text=Por%20si militud%20de%20uso%2C%20se,especial%20del%20aceite%20de%20oliva.

Espinoza, M., Zafimahova, A., Maldonado, P., Ducreucq, E., Poncet, C.(2015). Grape seed and apple tannins: Emulsifying and antioxidant properties. Food Chemistry, 178, 38–44. https://doi.org/10.1016/j.foodchem.2015.01.056

Errecalde, J. (2004). ¿Cuáles son los mecanismos de acción de los antibióticos?. Recuperado de www.fao.org/3/y5468s/y5468s05.htm

Food and Agriculture Organization. (11 de Noviembre de 2017). Recetas con "desperdicios" de alimentos, entre las estrategias innovadoras presentadas en el foro internacional sobre pérdidas y desperdicios. www.fao.org/colombia/noticias/detail-events/es/c/1062710/#:~:text=Seg%C3%BAn%20el%20estudio%20realizado%20por,pierden%20y%2012%25%20se%20desperdician.

Gimeno, E. (2004). Compuestos fenólicos. Un análisis de sus beneficios para la salud. Ámbito Farmacéutico Nutrición, 23 (6), 80–84. www.elsevier.es/es-revista-offarm-4-articulo-compuestos-fenolicos-un-analisis-sus-13063508

González, D. (2017). Aprovechamiento de residuos agroindustriales en Colombia. Revista de Investigación Agraria y Ambiental, 8(2). 141–149. https://dialnet.unirioja.es/descarga/articulo/6285350.pdf

Guevara, A. Salomón, M. Oliveros, A. Guevara, E. Guevara, M. & Medina,Z. (2007). Sepsis por *Chromobacterium violaceum* pigmentado y no pigmentado. Revista Chil Infect., 24(5), 402–405. https://scielo.conicyt.cl/pdf/rci/v24n5/art10.pdf

Hashim, S., Salman, S., & Hassan, R. (2020). Study of the effect of the seed grape *Vitis vinifera* plant extract on some pathogenic bacteria and fungi. Journal of Global Pharma Technology, 12, 197–201. https://1library.net/document/zle9eklq-study-effect-grape-vinifera-plant-extract-pathogenic-bacteria.html

Hatamnia, A., Abbaspour, N., & Darvishzadeh, R. (2014). Antioxidant activity and phenolic profile of different parts of Bene (*Pistacia atlantica* subsp. *kurdica*) fruits. Food Chemistry, 145, 306–311. 10.1016/j.foodchem.2013.08.031

Hernández, J. Trujillo, Y. & Durán D. (2011). Contenido fenólico e identificación de levaduras de importancia vínica de la uva isabella (*Vitis labrusca*) procedente de villa del rosario (Norte de Santander). Revista de la Facultad de Química Farmacéutica. 18(1), 17–25. www.scielo.org.co/pdf/vitae/v18n1/v18n1a03.pdf

Herrera, M., Catarinella, G., Mora, D., Obando, C., Moya,T. (2005). *Chromobacterium violaceum* Sensibilidad Antimicrobiana. Revista Médica del Hospital Nacional de Niños, 40 (1). 05–08. www.scielo.sa.cr/pdf/rmhnn/v40n1/3566.pdf

Huerta, N. (2020). *Escherichia coli*. Una revisión bibliográfica. Ocronos Editorial Científico Técnica.    https://revistamedica.com/escherichia-coli-revision-bibliografica/#:~:text=INTRODUCCI%C3%93N-,DESCUBRIMIENTO%20E%20HISTORIA,las%20deposiciones%20de%20los%20ni%C3%B1os.&text=Por%20lo%20general%2C%20cepas%20diferentes%20de%20E.

Instituto de Hidrología, Meteorología y Estudios Ambientales – IDEAM. (2018). Fenol. http://documentacion.ideam.gov.co/openbiblio/bvirtual/018903/Links/Guia14.pdf

Lek, C., Sam, C., Yin, W., Tan, L., Krishnan, T., Chong, Y., & Chan, K. (2013). Plant-derived natural products as sources of anti-quorum sensing compounds. Sensors, 13(5), 6217–6228. https://doi.org/10.3390/s130506217

March, G., & Eiros, J. (2013). Quorum sensing en bacterias y levaduras. Medicina Clínica, 141(8), 353–357. doi: https://doi.org/10.1016/j.medcli.2013.02.031

Mejía, A., Herrera, B., Salazar, M., Rojas, F., Gavín, V., & Escobar, J. (2017) Tomillo (*Thymus vulgaris*) como agente antimicrobiano en la producción de fresco. Revista Amazónica Ciencia y Tecnología, 6 (1). 45–54. https://dialnet.unirioja.es/servlet/articulo?codigo=6145604

Miller, M., & Bassler, B.,(2001). Quorum sensing in bacteria. Microbiol, 55(1), 165–189. doi 10.1146/annurev.micro.55.1.165

Mustafa, A, & Turner, C., (2011). Pressurized liquid extraction as a green approach in food and herbal plants extraction: A review. Analytica Chimica Acta, 703(1), 8–18. https://doi.org/10.1016/j.aca.2011.07.018

Navarro, A., Carmona, J., Font, X., (1996). Contaminación de suelos y aguas subterráneas por vertidos industriales. Acta Geologica Hispanica, 30 (1–3), 49–62. http://revistes.ub.edu/index.php/ActaGeologica/article/viewFile/4562/5814

Ng, K., Lyu, X., Mark, R., & Chen, W. (2019). Antimicrobial and antioxidant activities of phenolic metabolites from flavonoid-producing yeast: Potential as natural food preservatives. Food Chemistry, 270, 123–129. https://doi.org/10.1016/j.foodchem.2018.07.077

Noguera, N., Ojeda, L., Jiménez, M., Kremisiky, M. (2017). Evaluación del potencial antibacteriano de extractos de semillas de cinco frutas tropicales. Revista venezolana de Ciencia y Tecnología de Alimentos, 8 (1), 033–044. www.researchgate.net/publication/316975594_Evaluacion_del_potencial_antibacteriano_de_extractos_de_semillas_de_cinco_frutas_tropicales_Evaluation_of_the_antibacterial_potential_of_extracts_from_seeds_of_five_tropical_fruits

Phaissa, A. (2009). Microbiología y Patogenia Staphylococcus aureus. Infecciones producidas por *Staphylococcus aureus*. Barcelona, España: ICG Marge, SL. https://books.google.com.co/books?hl=es&lr=&id=qFRukXHQX6QC&oi=fnd&pg=PA9&dq=staphylococcus+aureus+historia&ots=ymkPSpzacp&sig=km5uBVtIJx7dev5PXTUIIzFV6SI0#v=onepage&q=staphylococcus%20aureus%20historia&f=false

Picazo, J. (2000). Sociedad Española de Enfermedades Infecciosas "Métodos básicos para el estudio de la sensibilidad a los antimicrobianos". www.seimc.org/contenidos/documentoscientificos/procedimientosmicrobiologia/seimc-procedimientomicrobiologia11.pdf

Pinilla, J. (12 de Abril de 2016). Colombia Duplicaría su Producción de Uva y el Valle del Cauca es el Líder. www.agronegocios.co/agricultura/colombia-duplicaria-su-produc cion-de-uva-y-el-valle-del-cauca-es-el-lider-2621888

Rodriguez, G. (2002).Principales características y diagnóstico de los grupos patógenos de Escherichia coli. Salud Pública Mex, 44 (5), 464–475. www.scielo.org.mx/scielo. php?script=sci_arttext&pid=S0036-36342002000500011

Rodriguez, M., Fraguela, J., Gonzalez, G., Muñoz, E., Carral, L. (2009). Evaluación del impacto ambiental provocado por las pinturas anti incrustantes utilizadas en las embarcaciones de recreo en los puertos deportivos de Galicia (Trabajo de Grado). España. www.resea rchgate.net/publication/238682671_Evaluacion_del_impacto_ambiental_provocado_ por_las_pinturas_antiincrustantes_utilizadas_en_las_embarcaciones_de_recreo_en_ los_puertos_deportivos_de_Galicia_Espana.

Semana magazine. (1 de Marzo de 2020). El 78% de los hogares colombianos no recicla. https://sostenibilidad.semana.com/medio-ambiente/articulo/el-78-de-los-hogares-colo mbianos-no-recicla/44231

Shenoy, S., Baliga, S., Wilson, G., & Kamath, N. (2002). *Chromobacterium violaceum* septi-cemia. Indian Journal of Practical Pediatrics, 69(4), 363–364. https://doi.org/10.1007/ BF02723225

Sheng, L,. Olsen, S,. Hu, J., Yue, W., Means, W,. Zhu, M. (2016). Inhibitory effects of grape seed extract on growth, quorum sensing, and virulence factors of CDC "top-six" non-O157 Shiga toxin producing *E. coli*. International Journal of Food Microbiology, 1–28. 10.1016/j.ijfoodmicro.2016.04.001

Silpa, K., Lisa, Y., Bhada-Tata, P., & Van, F,. (2018) . What a Waste 2.0: A Global Snapshot of Solid Waste Management to 2050. World Bank Group. Washington, DC. https://openkn owledge.worldbank.org/handle/10986/30317

Silván, J., Mingo, E., Hidalgo, M., Pascual, S., Carrascosa, A., Martinez, A. (2013). Antibacterial activity of a grape seed extract and its fractions against *Campylobacter* spp. Food Control, 29 (1), 25–31. https://doi.org/10.1016/j.foodcont.2012.05.063

State Registry of Emissions and Polluting Sources (Registro estatal de emisiones y fuentes contaminantes). (2007). ¿Qué son los fenoles? www.prtr-es.es/Fenoles, 15658,11,2007.html.

Tian, Y., Yingsa,W., Ma, Y., Zhu, P,. He, J., & Lei, J. (2017). Optimization of subcritical water extraction of resveratrol from grape seeds by response surface methodology. Applied Sciences, 7 (321), 1–12. 10.3390/app7040321

Viola, C., Alberto, M., Cartagena, E., & Arena, M., (2020). Inhibición de motilidad y Quorum sensing bacteriano por desechos de vinificación. Revista Iberoamericana Interdisciplinar de Métodos, Modelización y Simulación. 12, 151–166. https://doi.org/10.46583/nereis_ 2020.12.580

Yada, S., Kamalesh, B., Sonwane, S., Guptha, I., Swetha, R. (2015) Quorum sensing inhibition, relevance to periodontics. Journal List, 7(1), 67–69. www.ncbi.nlm.nih.gov/pmc/artic les/PMC4336667/

Yang, C., & Li, Y. (2011). *Chromobacterium violaceum* infection: A clinical review of an important but neglected infection. Journal of the Chinese Medical Association, 74(10), 435–441. https://doi.org/10.1016/j.jcma.2011.08.013

Yang, L., Qu, H., Mao, G., Zhao, T., Li, F., Zhu, B., Zhang, B., Wu, X. (2013). Optimization of subcritical water extraction of polysaccharides from grifola frondosa using response sur-face methodology. Revista Pharmacognosy, 9 (34), 120–129. 10.4103/0973-1296.111262

# Index

For Product Safety Concerns and Information please contact our EU
representative  GPSR@taylorandfrancis.com
Taylor & Francis Verlag GmbH, Kaufingerstraße 24, 80331 München, Germany

www.ingramcontent.com/pod-product-compliance
Lightning Source LLC
Chambersburg PA
CBHW070736220326
41598CB00024BA/3447